T0229854

R Visualizations

R Visualizations
Derive Meaning from Data

David W. Gerbing

The School of Business
Portland State University

CRC Press
Taylor & Francis Group
Boca Raton London New York

CRC Press is an imprint of the
Taylor & Francis Group, an **informa** business

A CHAPMAN & HALL BOOK

First edition published 2020
by CRC Press
6000 Broken Sound Parkway NW, Suite 300, Boca Raton, FL 33487-2742

and by CRC Press
2 Park Square, Milton Park, Abingdon, Oxon, OX14 4RN

Library of Congress Cataloging-in-Publication Data

Names: Gerbing, David W., author.
Title: R visualizations : derive meaning from data / David W. Gerbing, The School of Business, Portland State University.
Description: 1st. | Boca Raton : CRC Press, 2020. | Includes bibliographical references and index.
Identifiers: LCCN 2020004865 | ISBN 9781138599635 (hardback) |
ISBN 9780429470837 (ebook)
Subjects: LCSH: Information visualization. | R (Computer program language)
Classification: LCC QA76.9.I52 G47 2020 | DDC 001.4/226--dc23
LC record available at https://lccn.loc.gov/2020004865

ISBN: 978-1-138-59963-5 (hbk)
ISBN: 978-0-429-47083-7 (ebk)

Typeset in LMRoman
by Nova Techset Private Limited, Bengaluru & Chennai, India

Contents

Preface

This book shows how to do data visualization using the R data analysis language. The emphasis is on *how to*, with examples from many R packages across a wide range of visualizations. The book features both a review of the highly capable `ggplot2` visualization package, as well as a rethinking of how to do data visualization with the author's `lessR` package.

Hadley Wickham's `ggplot2` package is the most downloaded R package, and has become the standard for R visualizations. A literal grammar of graphics, the many `ggplot2` functions can create an impressive range of visualizations. By sequencing a chain of functions, build a visualization layer by layer with a virtually endless expression of possibilities.

The `lessR` package developed from a complementary perspective: Imagine the simplest code possible to generate data visualizations. First, minimize the number of functions. Whereas, for example, `ggplot2` requires three functions to build a basic bar chart, `lessR` across a wide range of visualizations only offers three core functions – `BarChart()`, `Histogram()`, and `Plot()` – plus `getColors()` to generate optional color palettes and `style()` to modify style themes. Obtain control of a visualization via the parameters of one of the three core functions with instructions, now all conveniently located in one place, the help manual for the respective function.

What is the most uncomplicated function call possible to generate a visualization? At a minimum, we need the function name and at least the name of one variable for which to transform the data values into an image. The variable usually, though not necessarily, exists within a table of data. The `lessR` visualization functions recognize a default name for the data table, the simplest name possible, *d* for data. To create a visualization of one or more variables in the *d* data table requires only entering into the R console one of the core visualization functions with the variable name(s) within the parentheses, for example, `Histogram(Salary)`.

The resulting `lessR` visualization is created from a top-down perspective instead of the usual bottom-up perspective, such as with `ggplot2`. Instead of building layers up one at a time, the `lessR` visualization by default includes the layers considered most visually appealing and useful. A layer not desired is removed instead of added. Again, simplify, and create the visualization with as little work as possible.

Further, the `lessR` visualization functions do more than visualizations. Their output includes a corresponding statistical analysis as well. `BarChart()` provides the frequency distribution and chi-square test. `Histogram()` provides the frequency distribution, outlier analysis, and summary statistics. `Plot()`, for example, when applied to two continuous variables, yields a descriptive and inferential analysis of the corresponding correlation coefficient.

The development of `lessR` began in 2009. Over a decade later, there are many influences on its progress. First, of course, is the massive amount of work by the R team and package developers to bring us their open-source vision of data analysis software. Also, I would like to thank Carlos Martin-Vide, of Rovira i Virgili University, Tarragona, Spain, for inviting me to participate in several sessions of the International School on Big Data. Presenting the seminars provided further incentive to develop `lessR`, as well as serve as a source of stimulation for perceptive feedback from students and other faculty. In particular, I am grateful for the input from Paul Bliese, of the University of South Carolina, Columbia, South Carolina, USA, who recommended the use of Deepayan Sarkar's lattice package as a basis for Trellis graphics upon which `lessR` now relies.

My own university home, The School of Business, Portland State University, provided a supportive environment where `lessR` could be written and where reside the many students who used successive versions of this software from its beginning. My students are not programmers and not statisticians. Most of them have never before seen a command-line environment. They do need, however, a means by which to analyze data and generate visualizations that communicate. How to introduce them to the power of R? The answer has been the development of `lessR`.

My wife, Rachel Maculan Sodré, has been my companion throughout. Would I otherwise have completed the project? Probably. Would it have been as much fun? Certainly not.

Chapter 1

Visualize Data

1.1   Introduction

1.1.1   Visualization and Analytics

data: Measurements of characteristics that describe different people, organizations, places, things, or events.

We are wonderfully competent visual processors. As we move about our daily life, we do what our ancestors back through the distant past did so well: Effortlessly process a panorama of shapes and images that surround us, patterns immersed within the landscape of our visual world. Modern life, however, delivers a new invention for us to consider: *data*. With data, we search for patterns such as normality, trends, and relationships, and we search for exceptions from these patterns. Examine rows and columns of data to uncover this information? Our distant ancestors never encountered tables of data, so our brains never adapted to evaluate data directly.

data visualization: Transform data values into visual objects with easily discernible attributes, such as shape, size, color, and location.

The solution? We return to our familiar form: visual images. To *visualize data*, we use computer technology to transform rows and columns of data into visible objects. We perceive these objects according to their *visual aesthetics*:

visual aesthetics: Visually perceived properties of objects.

- ▷ as different shapes (points, lines, bars)
- ▷ displayed at varying sizes (areas, lengths)
- ▷ with a palette of different colors (hue, saturation, brightness, transparency)
- ▷ which occupy different positions (by axes that define a coordinate system)

Visual aesthetics focus our perception on emergent patterns inherent in the data. We literally see the distributions and relationships. The instructions for creating these visualizations found throughout this book provide pleasing aesthetics by default. The resulting visualizations may also be customized to further accentuate the visible patterns and relationships.

data analytics: Clean and organize data, visualize and summarize data, and build predictive models.

Modern *data analytics* begins with organizing and preparing data. Then apply data visualization and other analytic methods to discover patterns, and build predictive models from those patterns. We apply these methods with computers, as outlined in Figure 1.1, to both traditional small data sets and big data.

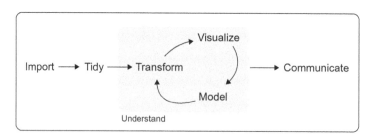

Figure 1.1: Overview of steps of data processing in an analysis project from Wickham and Garrett (2017).

data science: Application of a wide range of data analytic procedures to realize insight and prediction.

The emerging field of *data science*, which has become a growth engine for jobs, demonstrates the viability of this analytic framework. The data scientist requires knowledge regarding multiple domains: the content area of the application, the relevant statistical techniques and visualizations, the use of the computer to implement the techniques, and the interpretation and implementation of the results. This book focuses on the meaning of the data visualizations and how to create them.

After gathering the data, read the data values from an external data file into the analysis system, or read the data from multiple sources and then integrate into a unified data set. Then clean and prepare the data, ready for analysis in the form of what is called *tidy data*. This data preparation step, accomplished with its own set of analytic tools, can devour much more time than the analysis itself. Too frequently, structural problems with the data's organization, miscoding, a variety of inconsistencies, missing data, and other issues hinder data preparation. To transform data from its initial version to a cleaned, tidied version amenable for analysis typically requires the analyst to *wrangle* with the data.

Data preparation may include transformation of variables into new variables. Transformations may be as simple as converting multiple measures of length for different variables into common units such as centimeters. Or, dichotomize a continuous variable into the lowest half of the values and the highest half of the values. Or, square the values of a variable to conform to a specified statistical distribution.

From a cleaned, organized data set, the analysis proceeds with two primary tools: visualization and modeling. The construction and interpretation of models is a central concept of statistical analysis. Models both predict and explain the values of one variable given the values of one or more other variables called the predictor or explanatory variables. The application of statistical models ranges from traditional multiple regression analysis to advanced, computationally intensive machine learning procedures. Visualization and modeling iteratively follow each other in succession such that each analysis informs revisions for the other.

This book shows how to create many visualizations of data, applicable throughout the analytic process, from a basic understanding of the data through the communication of results to the intended audience.

▷ During the discovery phase of the analysis, explore data to gain insights into underlying patterns such as relationships and trends.

▷ After the analysis, communicate the results to the intended and possibly non-technical audience.

Data visualization facilitates both discovery and communication.

1.1.2 Open-Source Software for Data Visualization

We power the computations of data analysis with functions from the software system R (R Core Team, 2019). R provides *functions* to accomplish each of the analytic processes outlined in Figure 1.1: data preparation and transformation, modeling, and, of course, visualization. R is as robust and comprehensive as any data analytic software. R is also open-source, free as in costs $0.00, and runs identically on Windows, Macintosh, and the free Linux/Unix operating systems. Always better to have faster computers with much RAM, but with moderate size data sets R runs fine on older, less powerful computers as well.

Read() function, Section 1.2.4, p. 12

tidy data table, Section 1.2.2, p. 7

wrangle: To herd, to round-up, to brawl.

data wrangling: The process of cleaning, tidying, and otherwise preparing data for analysis.

goal: Discover then communicate usable information from underlying patterns inherent in data.

R: Comprehensive, open-source, cross-platform software for data analytics.

function: Procedure to accomplish a specific, repeatable task, here a specific statistical set of computations.

Simple visualization function calls

bar chart, Figure 2.1, p. 29

Although R offers of a sophisticated programming language for data analysis, obtain most of the visualizations presented in this book with one simple R function call. For example, consider the function `BarChart()`, explained in detail in the following two chapters. Given the data, create a colorful, labeled bar chart and corresponding statistical analysis of a variable called *Dept* with the following function call.

```
BarChart(Dept)
```

Once the data are prepared for analysis, the skills required for many of the simple R function calls demonstrated throughout this book require no R programming, no R expertise.

As discussed, data wrangling typically consumes most of the time and effort required for a data analysis. The software expertise to accomplish this preparation also consumes most of the time and effort to learn a data analysis computer language. Fortunately, the analyst can maximize expertise in the data analysis language of choice for data wrangling and still access the visualization functions presented in this book.

Python: Comprehensive, open-source, cross-platform software for data analytics plus general programming.

For example, the other primary open-source, free data analytic software system is *Python*, which provides data wrangling and other capabilities on par with R. Data wrangle the original data with the language of choice, such as R, Python, or whatever. Then easily write the cleaned data in a widely understood file format such as Excel or `csv`. Then, from the R environment, read the data. Or, access R functions directly from Python with the `rpy2` package.

write data, Section 1.2.7, p. 22

Visualize with simple function calls, such as `BarChart(Dept)`. Throughout this book find many examples of a one statement read function call and a one statement visualization function call. The choice of language for data wrangling is irrelevant. All that are needed is to access the R environment, read the cleaned data, and then access a simple visualization function, as shown throughout this book.

Get and Run R

Download and install R from `cloud.r-project.org`. From the links displayed at the top of the resulting web page, select your operating system.

- *Windows*: Click on the link toward the top of the obtained page, <u>base</u>. At the top of the following page, click on the link to the current version of R for the Windows operating system, such as <u>Download R 4.0.0 for Windows</u>.

- *Mac*: Several paragraphs down the page, click the link for the current version of R, such as <u>R-4.0.0.pkg</u>.

Once downloaded, the installation proceeds as with any app. Accept the given defaults for each step of the process. When installed, run the R app, such as by double-clicking on the application's icon in your file system display.

When opened, R displays a window called the *console*, which presents a prompt, >
as Figure 1.2 illustrates. Enter instructions into the R console at the > prompt. The
space to enter information is the *command line*.

> Console

> >

Figure 1.2: The command line with the R prompt within the console window.

Many, perhaps most, analysts choose to access the R console from within another
free application called RStudio (Team, 2016). Download the free desktop version at:

https://www.rstudio.com/products/rstudio/download/

RStudio enhances the environment for R analysis. Run the RStudio app to obtain
the same R console as from the standard R app, but as a windowpane. RStudio
divides the full window into up to four distinct panes that each provide useful
information. In addition to the console, panes can contain information such as one
or more R code files, a history of R commands entered in the current and previous
sessions, and visualizations.

Contributed packages

R organizes related functions into a group called a *package*. To install R is to install a
set of downloaded packages, such as the `graphics` package. Together these packages
form *base R*, developed and maintained by the R Core Team (2019). Beginning an
R session loads into working memory the installed base R packages. R users access
these functions when running R.

One of the strengths of R is extensibility. Anyone can write R functions, and,
once they pass a series of stringent tests, upload them to the official R servers, the
CRAN network, as a *contributed package*. The functions in a contributed package
supplement the large number of already available data analysis and other base
R functions. The functions from contributed packages work within the standard
R environment. This book presents the widely used `ggplot2` package for data
visualization, as well as the easier to use `lessR` package for both data visualization
and general data analytics.

To access the functions within a contributed package, first download the package from
a chosen `CRAN` server. Hadley Wickham, the author of `ggplot2`, and his colleagues
wrote several packages designed to work together. Their package simultaneously
downloads `ggplot2` and related packages called the `tidyverse`. To download and
install contributed packages, invoke the function `install.packages()` at the R
command prompt, >.

R Input *Download and install contributed visualization packages*

```
install.packages("tidyverse")
install.packages("lessR")
```

R console: Enter
commands into this
window and receive
text output.

R command line:
The place to enter
commands in
response to R's
command prompt, >.

RStudio: An
application that
provides an enhanced
environment for
running R.

R package: A
group of related
functions.

base R: The set of
R packages that
define an initially
installed R system.

**contributed
package**: An R
package provided by
one or more R users.

install.packages()
base R function:
Download packages
from the R servers.

R provides a list of CRAN servers around the world from which to select the desired server to download the packages.

R library:
Directory (folder) on
your computer that
stores the R packages.

R organizes the packages on your computer into a directory (folder) that R calls the *library*. To access the functions, data, and other information from a contributed package, invoke the `library()` function, which retrieves the package from the library.

library() function:
Access functions from
contributed packages.

> **R Input** *Provide for directly accessing functions from contributed packages*
>
> ```
> library("tidyverse")
> library("lessR")
> ```

Begin each R session with the `library()` function to access the functions and data from each relevant package. For data visualization, from the perspective of this book, that often means access to the tidyverse, `lessR`, or both.

Keep the set of installed packages current by occasionally re-downloading and re-installing more recent versions.

> **R Input** *Update installed packages*
>
> ```
> update.packages(ask=FALSE)
> ```

The optional `update.packages()` parameter `ask` set to `FALSE` informs R to not prompt for the update of each package for which an update exists. During the update process, you may also be asked:

```
    Do you want to install from sources the packages
which need compilation?
        (Yes/no/cancel)
```

If you do not know what that means, then answer **no**.

1.2   Data

Data analysis begins, naturally enough, with data, the measured data values. Part of the data preparation process organizes data into a rectangular table. The result is called a standard data table or, when stored within R, an R object called a *data frame*.

data frame: A
container that stores
a table of data values.

1.2.1   R Objects

Working with R is working with R objects such as variables and data frames. Store and access the R objects primarily through the R *workspace*, also called the user's workspace or *global environment*. Enter instructions at the R console's command prompt, >, to create objects.

workspace: The
environment where R
objects called data
structures typically
reside.

To enter data into an R object, use the equal sign, =. Or, use the more informative *assignment operator*, <-, a "less than sign" followed by a "minus sign". The arrowhead indicates the direction of the assignment. The assignment operator, or equals sign, relates two expressions. The value on the left of <- names the object that stores the data values assigned from the expression on the right.

To illustrate, begin with the simplest object, which merely holds a single value such as a number. Enter an expression at the R console such as the following at the command prompt:

```
x <- 5
```

R defines the object x in the workspace, and then assigns the value of 5 to the object, available for future analysis.

To display the contents of any R object at the console, list the name of the object at the command prompt, an implicit call to the R function `print()`. An example of the display of this simple object x, excerpted from an R session, appears in Listing 1.1.

```
> x <- 5
> x
[1] 5
```

Listing 1.1: Display the contents of the object named x.

Find the R output on the line that follows the entered name of the object, x. The `[1]` displayed in the R output indicates that this line of output begins with the first element of the displayed R object. Our object is only a single number, so entering its name at the console only displays one element, the number 5. The next section of this chapter introduces the R objects that we care most about, the data table.

The same principles that apply to a simple value such as setting x to 5 generalize to more complex R objects such as data frames. Create R visualizations from functions that access data stored in an R object that contains an entire data table with many data values.

The data table imposes a strict structure on the organization of the data. Each column of the data table contains the data values for a variable; all data values entered with a consistent format. Each row of the data table contains the data values for a single observation across all the variables in the data table, such as for a single person, company, or geographical region. A data table that satisfies these constraints is called a *tidy data table*.

1.2.2 Employee Data Example

The human resources department of a company recorded the following information for each employee: name, years of employment, gender, department employed, annual salary, job satisfaction with the work environment, health plan, and a score on a pre-test followed by classroom instruction, and then the corresponding score on

assignment statement: <- assigns the value of an expression on the right to a variable or object on the left.

example data table, Section 1.2.2, p. 7

data analytic process, Figure 1.1, p. 2

tidy data: Table of data with all values for a single person, company, etc. within a row, and all consistently formatted values for each variable within a column.

data file: A file on a computer system that contains data, usually organized into a table of rows and columns.

a post-test. Locate a *data file* on the web in Excel format with these measurements organized into a table for 37 employees.

> `http://lessRstats.com/data/employee.xlsx`

Figure 1.3 displays the first nine rows of the Excel version of this data file, with the variable names in the first row, the usual format.

	A	B	C	D	E	F	G	H	I
1	Name	Years	Gender	Dept	Salary	Satisfaction	HealthPlan	Pre	Post
2	Ritchie, Darnell	7	M	ADMN	$43,788.26	med	1	82	92
3	Wu, James		M	SALE	$84,494.58	low	1	62	74
4	Hoang, Binh	15	M	SALE	$101,074.86	low	3	96	97
5	Jones, Alissa	5	F		$43,772.58		1	65	62
6	Downs, Deborah	7	F	FINC	$47,139.90	high	2	90	86
7	Afshari, Anbar	6	F	ADMN	$59,441.93	high	2	100	100
8	Knox, Michael	18	M	MKTG	$89,062.66	med	3	81	84
9	Campagna, Justin	8	M	SALE	$62,321.36	low	1	76	84

Figure 1.3: Variable names in the first row, followed by the first eight rows of data in this tidy employee data table stored in an Excel worksheet.

data value: Value of a single measurement or classification.

unit of analysis: Class of people, organizations, things or places that are measured.

Each cell of this table represents a single *data value*. For example, the person listed in the first row of data, the second row of the worksheet, is Darnell Ritchie. The data value for his annual *Salary* is $43788.26. An essential characteristic of a data table is the *unit of analysis*, the object of study. In this example, the unit of analysis is a person, an employee at a specific company.

variable: Attribute that varies from unit to unit (e.g., different people.)

variable name: Concise name that identifies a variable for analysis.

A *variable* is a characteristic of an object or event with different values for different instances of the unit of analysis, different people, organizations, etc. Reference the variable in any subsequent data analysis by its concise *variable name*, usually less than about 10 characters. The first row of the data table usually contains the variable names. The data values for each variable lie within the same column. The data values of the first column in this particular data table list the unique ID values that identify the employee for the corresponding row of data values. All the data values in a single row, collected under the same conditions, such as the same experimental treatment, together are called an *observation*. For example, the first observation, the second row of the Excel worksheet, consists of all the data values for Darnell Ritchie.

observation: Data values for a single instance of the unit of the analysis, such as a person or object collected under the same circumstances.

There are many ways to represent data that is not tidy. One example would be transposing (rotating) the data table such as in Figure 1.3 so that the variables are in rows instead of columns. Or, a summary of data values such as a list of how often each data value occurred, its distribution, is also not tidy. Or, consider a data table with a variable *Gender*. If some values for Male are coded as `"M"`, other values coded as `"m"`, and other values coded as `"Male"`, then the consistency of the format is violated, and the data table is not tidy.

Much of the work of data analytics tidies the data, to arrange the data values in a format amenable to analysis. Data analysis software, including base R, includes tools to transform and re-arrange data to become tidy. Many base R functions are available for tidying data. Also available are functions, written by Hadley Wickham and colleagues, in the `tidyr` and `dplyr` packages, part of the related set of packages called the `tidyverse`.

tidyverse, Section 1.1.2, p. 5

1.2.3 Types of Variables

When analyzing data, always distinguish between continuous variables and categorical variables, the distinction between a variable defined on a numerical scale vs. one defined as a set of non-numerical categories. Two issues here. First is the intrinsic definition of a variable apart from any computer program. Second is how to represent these different types of variables as data structures stored within the computer's memory, the *data storage type*.

data types: The format of how the data values for a variable are stored in the computer.

Continuous variables

The theoretical values for *continuous variables* are ordered along a quantitative continuum, the abstraction of the infinitely dense real number line, in which an unlimited number of numeric values lie between any two values. Examples of continuous variables for a person are: Age, Salary, or extent of Agreement with an opinion about some political issue. Continuous variables that characterize a car include MPG, and Weight; and for a light bulb, Mean Number of Hours until Failure, and Electrical Consumption per Hour (kilowatt hours). Also refer to a continuous variable as a *quantitative variable*.

continuous variable: A variable with numerical values ordered on a continuum.

Measurement *always* organizes the resulting data values into discrete categories, even for continuous variables. The value of the variable as it exists differs from its measurement for two reasons. First, the corresponding data value never attains sufficient precision. Nothing, for example, weighs exactly 2 lbs., or 2.01 lbs. or even 2.0000000001 lbs. Instead of the exact value, express a measurement of weight to an *acceptable* degree of precision, such as the nearest pound. Second, measurement can introduce measurement error, the measured value placed in the wrong category, such as incorrectly reading the weight on a bathroom scale as 155 lbs. instead of 156 lbs.

Store the data on a computer system as a data file. Then read the data into an analysis system such as R. Distinguish between four distinct representations of the values of a variable:

▷ Actual values as they exist apart from their measurement

▷ Data values as recorded measurements of the actual values

▷ Data values stored within a computer file

▷ Data values within the analysis system such as R

The most general representation of continuous variables within an R analysis is the R data type *double*, which represents numbers stored with decimal digits. The term `double` applied to computer usage refers to the amount of memory allocated to store the numeric value, in this case, what is called "double precision" for 64 bits

double: Storage type for numeric values with decimal digits.

(binary integers or on-off switches) per data value. For example, R stores the data values for the variable *Salary* in the Employee Data table from Figure 1.3 as type `double`.

Employee Data table,
Figure 1.3, p. 8

Another R data storage type for numeric variables is type *integer*, for numbers without decimal digits. Although not strictly needed, an actual integer value stored as type `integer` requires less storage space than does the corresponding double precision value. Another benefit is that the integer value is stored as an exact value. Computers do not store numbers as decimal digits, but rather as binary digits, and often the binary representation of a double precision number is not the same to many decimal digits as is its decimal representation. An example of a variable stored as type `integer` is the number of Years employed from the Employee Data table from Figure 1.3.

integer: Storage type for numeric values with no decimal digits.

Apart from computer storage, conceptually interpret data values according to a numeric scale as one of two types. *Ratio variables* follow a numeric scale with strong numeric properties. There is a fixed zero point, and values on either side of zero proportionality scale. In particular, compare two different values by their ratios, such that, for example, 20 is twice the magnitude as 10. Also, equal intervals of magnitude separate values that are equal distance from each other. For example, the distance between 21 and 22 represents the same underlying magnitude of difference for 22 and 23.

ratio variable: Data for a variable that defines a numerical scale with fixed zero point and equal intervals.

A weaker numerical scale describes *interval data*, which maintains the equal interval property of ratio data but does not have a fixed, natural zero point. As an example, compare Fahrenheit and Celsius temperatures. Each temperature scale, for example, has a different value of zero regarding the magnitude of the temperature. 0°F is not the same temperature as 0°C. Ratio comparisons are not valid; 20°F is *not* twice as warm as 10°F.

interval variable: Data for a variable defines a numerical scale without a fixed zero point but equal intervals.

Categorical variables

In contrast to continuous variables, the values of a *categorical variable* consist of a relatively small set of non-numeric categories called levels. One origin of a categorical variable is a continuous variable measured so imprecisely that instead of a numerical scale, the values are represented with only a small number of categories. Suppose hospital personnel quickly classify people admitted to the emergency room into one of only three types according to the severity of their injury: Low, Medium, and High. The underlying variable regarding *Severity* is continuous. This simple rating scale recognizes that some injuries are more severe than others but limits the measurements to only one of three non-numeric categories. Data values for a variable grouped into ordered, non-numeric categories, rankings, represent an *ordinal variable*.

categorical variable: A variable with non-numeric categories as values.

The distances on the continuum of the magnitude of severity between adjacent categories are not assumed equal. Medium Severity of Injury could be closer to Low Severity than to High Severity of Injury. Another example is a ranking, such as the outcome of a race with the winners ranked in order of finish: 1st, 2nd, and 3rd. The finish times of each contestant represent a continuous variable. Merely ranking

ordinal variable: Ordered categories, rankings.

contestants by order of their finish, however, does not communicate if the race was extremely close or if the winner finished well ahead of the nearest competitor.

The other context for a categorical variable applies to unordered categories. For example, a preference for red, rose, or white wine results in unordered categories. Other examples of categorical variables are Cola Preference, State of Residence in the USA, or Football Jersey Number. The number on the jersey consists of numeric digits, but in this context, those digits represent characters instead of numerical digits. A categorical variable is sometimes also called a *qualitative variable*.

Classification of the data values of a categorical variable into discrete, unordered categories yields a *nominal variable*. An example is Male, Female, and Other as the values of *Gender*. There is no underlying ordering, only discrete values.

nominal variable: Data values for a variable grouped into unordered categories.
character variable: A storage type for non-numeric values stored as character strings.

One storage type reserved for non-numeric data values is a variable of type `character`, which represents character values as literal character strings. A second, more useful storage type for non-numeric values is what R calls a `factor`, which stores the categories as integers but displays the categories as non-numeric labels, Male instead of 1 and Female instead of 2, for example. Other common variable types encode dates and times, such as variable type `Date`, which encodes calendar dates.[1]

factor, Section 1.2.6, p. 19

type Date, Section 7.2.4, p. 168

A potential confusion is the lack of a 1-to-1 correspondence between the type of variable, numeric or character, and the data storage type, numeric or character. The relatively small number of unique, non-numeric values of a categorical variable can correspond to any data type. For example, in the data table from Figure 1.3, the categorical variable *Plan* has three integer values – 1, 2, and 3 – that correspond to three health plans. Although the data values are numbers, they only serve as labels and could be replaced with any other set of arbitrary labels such as A, B and C.

To avoid confusion, represent categorical variables as variable type `character` or, preferably as type `factor`. No one confuses the values Male and Female, or the values M and F such as in Figure 1.3. However, for *Gender* stored as an integer variable, does the 0 represent Male, Female, or something else? As an additional safeguard, a variable of type `character` or of type `factor` have values that cannot be mistakenly treated as numeric values and then subjected to inappropriate numerical analysis. There is no mean of the values of M and F, but there is for values encoded as 0 and 1.

1.2.4 Read Data

The data resides in an external file. Data analysis begins by reading the data from the file into a data structure called a *data frame*[2]. For `lessR` analyses, typically name the data frame *d* for data, the default data frame name for the `lessR` data

data table, Figure 1.3, p. 8

---

[1]Less common storage types are also available, `complex` for complex numbers and `logical` for a categorical variable with only two values encoded as `TRUE` and `FALSE`.

[2]The tidyverse set of packages, including `ggplot2`, applies its own version of a data frame called a `tibble`. All `lessR` and `ggplot2` visualization functions process either the standard R data frame or the `tidyverse` tibble version of a data frame.

data frame: A data table within an R session, ready for analysis.

analytic and visualization functions. Unless the name of the input data frame differs from *d*, no need to explicitly invoke the **data** parameter to specify the name of the data frame.

The **lessR** Read() Function

Read(), lessR: Read data from a file into a data frame (table).

Base R and contributed packages provide multiple read functions for data stored in a variety of file formats. Two common types of file formats are text files, comma (csv) or tab-delimited (tsv), and Excel files, all of which by default list the variable names in the first row. The **lessR** function, **Read()**, abbreviated **rd()**, automatically selects a corresponding more specific read function by default according to file type, reads the data into a data frame, and then provides feedback regarding the information that was read. The same **Read()** function with the same syntax applies to many different file formats, freeing the analyst from loading the relevant packages and then explicitly accessing the corresponding more specific read function[3].

read employee data, Section 1.2.4, p. 12

In the following example, read the employee data from an external Excel data file from the web into the *d* data frame. The primary information passed to **Read()**, the only required information, identifies the file to be read as a character string enclosed in quotes that reveals the file's location. To read from a web file specify the **http** protocol as part of the file reference.

R Input *Read data from a file on the web into the* d *data frame*

```
lessR: d <- Read("http://lessRstats.com/data/employee.xlsx")
```

install.packages function, Section 1.1.2, p. 5

This data file, referenced throughout this book, is also included with **lessR**, available without accessing the Internet once **lessR** has been downloaded. To access data included with **lessR**, reference the file only with its name, here **"Employee"**.

R Input *Read data from an included* lessR *data set into the* d *data frame*

```
lessR: d <- Read("Employee")
```

browse for a file: Navigate your file system via the usual window provided by your operating system to locate and select a file.

An empty character string passed to **Read()**, **""**, instructs R for the user to *browse* interactively for the file from which to read the data.

R Input *Browse for data file to read data into the R data frame* d

```
lessR: d <- Read("")          or          > d <- rd("")
```

Read() then displays the full file reference (path name) at the console, which can be copied and pasted into a subsequent **Read()** statement. Save this statement for future reading of the same data, so that the analysis is reproducible.

To better understand the structure of the data read into a data frame, **Read()** by default lists its basic characteristics, illustrated in Listing 1.2 for these data. Verify the structure and content of the read data. Many things can go wrong with

---

[3]**Read()** invokes the base R function **read.csv()** to read comma-delimited or tab-delimited text files, and the **read.xlsx()** function from Alexander Walker's **openxlsx** package to read data tables stored in Excel format. **Read()** also reads files from the other data analysis systems **SAS** and **SPSS**, as well as native data files from R itself.

data. Perhaps some of the data values were incorrectly entered into the data file. Perhaps data values intended to be numerical were instead read into a character string variable. Perhaps there is too much missing data for some variables to permit meaningful analysis.

```
        Variable                 Missing Unique
            Name     Type Values  Values Values  First and last values
        ------------------------------------------------------------------
            Name character    37       0     37  Tian, Fang ... Jones, Alissa
           Years integer      36       1     16  7  NA  15 ... 1  2  10
          Gender character    37       0      2  M   M   M ... F  F  M
            Dept character    36       1      5  ADMN  SALE  ... SALE  FINC
          Salary double       37       0     37  43788.26  ...  47562.36
          JobSat character    35       2      3  med  low  ... low  high
            Plan integer      37       0      3  1  1  3 ... 2  2  1
             Pre integer      37       0     27  82  62  96 ... 83  59  80
            Post integer      37       0     22  92  74  97 ... 90  71  87
```

Listing 1.2: Read() output.

To turn off output to the R console, as with all `lessR` functions, set `quiet` to `TRUE`, either in the individual function call, or in a call to `style()` to implement for all successive analyses until changed again.

style(), `lessR` function, Section 10.3, p. 223

Subset data frames

Obtain precise control over the display of the data, or create new data frames, with the base R function **Extract**. This function is called not by its name, but, when applied to a data table, by one or more arguments separated by commas within square brackets. Specify the name of an R object, such as a data frame, followed by the brackets. Within the square brackets, before the comma specify an expression that defines the range of relevant rows, and after the comma, specify an expression that defines the range of relevant columns. To select all rows, enter nothing before the comma. To select all columns, enter nothing after the comma.

Extract function, **base** R: Expression that extracts a subset of a data table or other R object.

The expression can be a vector of row or column indices, a logical expression, or, for the columns, a set of variable names in quotes. The following example selects the first three rows and columns 4 through 7 of the *d* data frame by specifying the respective row and column indices. Because the selection in this example is not assigned to output by an assignment statement, R directs the selection to the R console for viewing.

```
d[1:3,4:7]
```

The subsetting operator has extensive use throughout the R ecosystem. Specify the vectors to define the row and column ranges in one of the many ways that vectors may be specified. For example, define a vector of integers of 1, 2, and 3 with the notation 1:3, as shown in Listing 1.3.

vector: A variable that consists of multiple values.

More generally, call the base R `c()` function to define a vector, such as `c(1,2,3)`. Vectors are frequently applied in R analyses, including visualization.

c(), base R function: Define a vector.

```
> 1:3
[1] 1 2 3
```

Listing 1.3: A vector of the first three integers.

The base R `Extract` function has several annoying limitations. If the expression for rows involves a variable in the data frame, the variable name must be preceded by the name of the data frame followed by the $, even though the data frame name has already been specified at the beginning of the call to `Extract`. For the columns, each variable name must be enclosed in quotes. Further, no variable ranges are allowed, instead enter the name of each variable in the list of variable names.

`.()`, `lessR`: index rows and columns within base R Extract

The `lessR` function, `.()`, addresses these limitations. Enclose the expression for the row filtering in `.()` to avoid re-listing the data frame name for every variable name in the expression. Enclose the selection of variables in `.()` to avoid having to enclose the names in quotes, and also to allow variable ranges. In the following example, the variables *Gender*, *Dept*, and *Salary* are contiguous in the data frame. To specify a range, list the first and last variables in the range, separated by a colon.

> **R Input** *Example of subsetting a data frame*
> *data*: d <- Read("http://lessRstats.com/data/employee.xlsx")
>
> *base R*: d[d$Salary > 95000, c("Gender", "Dept", "Salary", "Post")]
> *lessR*: d[.(Salary > 95000), .(Gender:Salary, Post)]

Both of these subsetting functions result in the identical output in Listing 1.4.

```
                Gender Dept    Salary Post
Hoang, Binh         M SALE 111074.86   97
Knox, Michael       M MKTG  99062.66   84
Sheppard, Cory      M FINC  95027.55   73
Correll, Trevon     M SALE 134419.23   94
James, Leslie       F ADMN 122563.38   70
Capelle, Adam       M ADMN 108138.43   81
```

Listing 1.4: A subset of the employee data frame of four variables for just those employees who make more than $95,000.

To enhance readability, the expression for rows can be set elsewhere to a character string `rows`, and the expression for columns set to a character string `cols`. Subset, then, with the literal expression `d[.(rows),.(cols)]`.

Both base R and the tidyverse provide other functions for filtering rows of data and selecting variables. However, the expression of the base R `Extract` function with the brackets `[ ]` in conjunction with `.()` provides a general and convenient expression for subsetting a data frame, without the need for additional functions.

Functions and their parameters

data analysis function: Procedure to access and analyze data.

A *data analysis function* applies statistical methods to process the data. To obtain the requested analysis, the output of the function, illustrated in Figure 1.4. R data

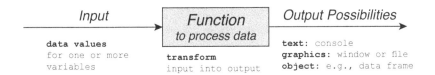

Figure 1.4: General procedure for functions that process data, either data analysis or data modification.

analysis functions process data, then output the results to some combination of the R console as text, to a graphics window or file as a visualization, and to an R object.

Whether working in R, Python, Excel, or most other computer languages, the definition of a function includes one or more *parameters*. The function performs its computations with a specific value of each parameter, the parameter's *argument*. Many, if not most, of the parameters typically have *default* values, invoked if not explicitly changed by the user. To call the function to perform its computations, specify the function name, and then within an opening and closing parentheses, specify relevant parameter values. Separate each parameter value by a comma.

parameter: Placeholder for a value supplied for a specific analysis with the function.

argument: Value of a parameter from which a function begins its computations.

For example, the first parameter value of the `Read()` function, `from`, specifies the location of the data file from which the function reads the data values into a data frame. There is no default value for this parameter, so specify a value to proceed. As illustrated in Figure 1.5, to explicitly name the `from` parameter in the function call, provide the parameter name, an equal sign, and then the provided value.

default: Given values of parameters made active unless the user overrides.

```
data      func-  para-
frame     tion   meter                          argument
 d   <-  Read  (from = "http://lessRstats.com/data/employee.xlsx")
```

Figure 1.5: Components of a function call.

Parameters with arguments listed in the same order in the definition of the function do not need to be named in the function call. Although the `from` parameter is explicitly named in Figure 1.5, it does not need to be because it is the first parameter in the definition of `Read()` and it is listed first.

View the parameter list of a function with the base R function `help()` with the function name provided as the first argument, as illustrated in Listing 1.5.

help(), base R function: Display information about the indicated function.

```
1  > help(Read)
2
3  Read(from=NULL, format=NULL, var_labels=FALSE)
```

Listing 1.5: Brief excerpt from the output of the `help()` function for `Read()`, which includes more parameters defined in the full listing and more information in general.

When a value is provided for a parameter in the definition of a function, then that value is the default value, the value invoked if not explicitly provided. For the `Read()` function, as shown in Listing 1.5, the parameter `var_labels` is set to `FALSE` when the function is invoked unless a value `TRUE` is explicitly provided. The value of the parameter `format` is initially set to `NULL`, which means that unless the value is provided by the user, the function computes its value from other provided information. In this example, `Read()` infers the file format from the filetype, such as an Excel file if the file type is `.xlsx`.

var_labels parameter,
Section 1.2.5, p. 18

Missing data

missing data: A
data value not
available in the data
table.

Another consideration of reading data is *missing data*. For example, the data value in the Excel file for Years of employment, the first variable, which occurs immediately after the name, is missing for the second row of data, the data for James Wu. Instead of a number for Years, there is only an empty cell. R indicates missing data values with `NA`, for Not Available. By default, when reading data R converts blank cells in Excel, or two adjacent commas in a `csv` file, to `NA`.

Excel data file,
Section 1.3, p. 8

**R missing data
code**: R codes
missing data values
as `NA`, not available.

In place of no information, literally a blank cell for Excel, or nothing entered between two commas for a `csv` formatted file, data may contain missing data codes. Examples include `-99`, to indicate missing numerical data, and `"XX"` for missing character data. Instruct `Read()` to process these values as missing with the parameter `missing`, as in the following example.

Missing data codes with `Read()`

```
lessR: d <- Read("", missing=c("XX",-99))
```

When this code is run, `Read()` replaces every `XX` and every `-99` that exists in the data file read into R with an `NA` in the resulting R data frame, here *d*.

Read() output,
Listing 1.2, p. 13

`lessR` presents a missing values analysis by variable as part of its default output to the R console. As seen in Listing 1.2, the column titled Missing Values provides counts of the number missing for each variable in the data table. Find the complementary information, the number of non-missing values, in the column Values. The values for each variable in these two columns should sum to the same number, here 37, the total number of rows in the data table.

Details() function,
`lessR`: Provide
details of specified
data frame, also brief
output with **db()**.

To obtain a missing values analysis by observation, that is, by row of the data table, invoke the `lessR` function `details()`, with default data frame name *d*. The unique output is shown in Listing 1.6. The information from `Read()` in Listing 1.2 that also is output by `details()` is not repeated in Listing 1.6.

For this data set there is little problem with missing data. The most missing data values for a variable is only two, for the variable *JobSat*. The most missing values for a row is for the fourth row of data, also with two missing values. For other data sets there may be large amounts of missing data for variables and/or rows of data, which may require further action, such as deleting incomplete rows or columns of data.

```
> details()

-------------------------------------------------------------
Dimensions: 9 variables over 37 rows of data

First two row names: 1       2
Last two row names:  36      37

...

Missing Data Analysis
-----------------------------------------------------
n.miss  Observation
1        2
2        4
1        31

Total number of cells in data table:   333
Total number of cells with the value missing:   4
-----------------------------------------------------
```

Listing 1.6: Partial output of lessR details() applied to default data frame d.

Read Fixed-Width Data

Another common data format is a text file with the data values for each variable defined by the columns in which they appear. Relying upon parameters included with the base R read functions, Read() can also read these columns. The illustration here is with the fixed-width 6-pt Likert response data for the original Machiavellianism scale, the 20-item Mach IV scale (Christie & Geis, 1970).

The responses for each of 351 persons to each of the 20 items – $m01$, $m02$, ..., $m20$ – were encoded according the following integers.

0 - Strongly Disagree
1 - Disagree
2 - Slightly Disagree
3 - Slightly Agree
4 - Agree
5 - Strongly Agree

The first three rows of data from this file follow, one row of data for the responses of each person. The first four columns are an ID field, the next column indicates *Gender* (0=Male, 1=Female), and then the integer responses to the 20 Mach IV items, one integer per column. Unlike the previous examples, a fixed-width data file does not have the variable names in the first row.

Data *First three lines of a data file with fixed-width columns*

```
010000415054154000401324
012700144033044011124431O
013412105440534140020240I
...
```

Use Read() to read these data values into the d data frame as follows.

> **R Input** *Read data from a file to the* d *data frame*
>
> ```
> lessR: d <- Read("http://lessRstats.com/data/Mach4.fwd",
>                  col.names=c("ID", "Gender", to("m",20)),
>                  widths=c(4,1,rep(1,20)))
> ```

To read a text file that does not have the variable names in the first row, apply the base R `col.names` parameter to define the names as part of the call to `Read()`. The base R `widths` parameter defines the respective widths of the columns, one column for the responses to each variable.

to(), lessR function: Construct a vector of variable names, numbered sequentially.

In this example, the `col.names` parameter indicates that the respective variable names are *ID*, *Gender*, and, as specified by the `lessR` function `to()`, *m01*, *m02*, to *m20*. The `widths` parameter specifies that the first variable occupies four columns, and that the next variable occupies a single column. The base R repetition function `rep()` specifies that the next 20 variables each occupy only a single column, short-hand for writing 20 1's, separated by commas.

1.2.5 Variable Labels

Variables with their data values are the fundamental unit of data analysis. Typically, we prefer brief variable names, which not only minimizes spelling errors but also lead to simpler input and more concise output. The names and values of many more variables can be listed in columns across a computer screen or printed with shorter variable names.

variable label: A phrase that describes a variable's meaning.

A short variable name, however, may not convey much information about the meaning of the corresponding variable. A *variable label* is a longer, more descriptive phrase that more fully describes the variable's meaning, such as "Annual Salary (USD)" instead of only Salary. Although variable labels are not part of the base R system, `lessR` allows for variable labels to label both text and visual output to aid the interpretation of results.

Enter the variable labels into a standard two-column worksheet in either Excel format or a `csv` text file, one column for the variable name and the second column for the corresponding label. Read this file with `Read()` into a data frame called `l`, setting the parameter `var_labels` to `TRUE`. The `lessR` analysis functions check for the existence of that data frame, and if it exists with variable labels for the corresponding variables in the analysis, the labels are used to enhance the text and visualization output. Figure 1.6 shows the file of variable names and labels for some of the variables in the Employee data set.

	A	B
1	Years	Time of Company Employment
2	Gender	Male or Female
3	Dept	Department Employed
4	Salary	Annual Salary (USD)

Figure 1.6: Excel file of variable names and variable labels, ready for import into a data frame.

Read the labels in the file into the l data frame (lowercase letter l for labels).

R Input *Read variable labels from the web into the l data frame*

```
l <- Read("http://lessRstats.com/data/employee_lbl.xlsx",
        var_labels=TRUE)
```

Any text output at the console, or within a visualization, by default appends the variable label to the variable name. For example, if present, in addition to the variable name, a variable label annotates an axis label for a visualization. Without the variable label, specify custom labels for the axes of a visualization with the base R parameters xlab and ylab.

xlab, ylab, base R parameters: Custom labels for the two axes of a visualization.

Variable labels can also be used with non-lessR analysis functions, such as with base R or ggplot2 functions. To do so, invoke the lessR function label(), called with the variable name. Replace the relevant character string, such as the specified label for the x-axis, with reference to label().

label, lessR function: Access variable labels for non-lessR functions.

1.2.6 Categorical Variables as Factors

R provides a variable type to represent categorical variables needed for data analysis and visualizations called a *factor*. Using the lessR function Read() or tidyverse functions such as readr() to read a data table into R reads variables with character string data values as type character. Of course, categorical variables with integer data values are read as numeric. Generally, convert categorical variables, read initially as integers or character strings, to factors before analysis begins.

factor: A variable with values of a categorical variable stored as integers that display as labels.

Categorical variables have a relatively small number of fixed categories (or levels). For example, three fixed responses to a survey item of Disagree, Neutral and Agree qualifies the associated variable for representation as an R factor. So does State of Residence in the USA with its 50 possibilities. Street addresses are represented as character strings, but would not be well represented as R factors because each employee likely resides at a different street address. There are as many (or almost as many) distinct addresses as there are people sampled for a data set, with no specified fixed set of addresses.

An R factor internally stores the levels of a categorical variable as integers but displays each level according to a label associated with the corresponding level, the *value label*. Represent a categorical variable as a factor to accomplish three primary tasks required for visualizations.

value label: Description of the level of a categorical value.

▷ Order the levels of the data values
▷ Attach meaningful labels to integer data values
▷ Provide levels that did not occur in the data

factor(), *base R*: Function to create an R factor variable.

A consideration of each follows. The basis for the conversion of a variable of type character or type integer to a factor is the base R factor() function. lessR provides a function, factors(), that extends the functionality of factor() and simplifies its use.

factors(), *lessR*: Function that creates one or more R factor variables.

Order levels

bar chart,
Section 3.1.1, p. 47

Three levels describe the categorical variable *JobSat* from the Employee data set: low, med and high. By default, R alphabetically orders the levels of a categorical variable, such as the bars in a bar chart. Visualizations that involve the *JobSat* variable unmodified present the levels in the wrong order: high, low, and med. To properly order, convert the variable from type `character` as initially read into a variable of type `factor`.

The base R `factor()` performs this transformation with parameter `levels` that specify the levels of the categorical variable of interest. List the levels in the desired order, what can be called their presentation order in subsequent visualizations. To specify the transformation directly, identify the corresponding data frame when referencing the *JobSat* variable so that R can locate the variable. Do so by preceding the variable name with the name of the data frame followed by a dollar sign, $.

```
base R: d$JobSat <- factor(d$JobSat, levels=c("low", "med", "high"))
```

Once converted, there is no need for the original character version of *JobSat*, so the preceding transformation replaced the original with the factor version. Or, create a new variable in the *d* frame by entering a new variable name in the left-hand side of the specification, before the `<-`.

Compared to the base R `factor()`, lessR `factors()` can be less verbose, and it generalizes from the application of a single variable to a vector of variables: Simultaneously convert to factors a sequential range of categorical variables, or any arbitrary vector of categorical variables.

Unlike `factor()`, which requires the specification of individual variables, `factors()` applies *only* to variables within a data frame, and returns the entire data frame as output, including the factor transformation. In this example, revise the *d* data frame with the resulting factor conversion. The default input data frame for lessR functions is *d*, so if no input data frame is specified with the `data` parameter, the specified variable(s) is assumed to be in the *d* data frame.

```
R Input Define a factor from a variable of type character
data: d <- Read("http://lessRstats.com/data/employee.csv")
```
```
d <- factors(JobSat, levels=c("low", "med", "high"))
```

In this example, replace the original *JobSat* variable with its factor version. Or, save the output from `factors()` into an entirely different data frame with all variables in the original data frame copied, including the newly created factor.

Ordinal variable,
Section 1.2.3, p. 10

The optional `ordered` parameter for `factor()`, and by extension, `factors()`, indicates that the levels progress in magnitude from "less" to "more". Ordering the levels with the `levels` parameter specifies their display order. Setting the `ordered` parameter to `TRUE` goes further than specifying the presentation order to indicate that the factor variable is an ordinal variable. By default, the value of `ordered` is FALSE, which indicates a nominal variable. For subsequent data visualizations, ordered factors have different default color palettes than non-ordered factors that reflect the underlying ordering.

Nominal variable,
Section 1.2.3, p. 11

Label integer values

Levels of a categorical variable may be coded as integers. For example, for the employee data, the variable *Plan* is categorical, coded in the data file with the integers 1, 2 and 3 that correspond to three health plans, respectively named GoodHealth, GetWell, and BestCare. The resulting visualizations are more meaningful with the output labeled with the names instead of integers, so transform a categorical variable read into an R data frame of type integer into a factor variable.

employee data,
Section 1.2.2, p. 7

Create a factor for one variable. Follow the same general procedure as the previous example that transforms a variable of type `character` into a `factor`, but also introduce another parameter called `labels` to provide more meaningful value labels.

```
R Input Define a factor from a variable of type integer
data: d <- Read("http://lessRstats.com/data/employee.csv")

d <- factors(Plan, levels=1:3,
           labels=c("GoodHealth", "GetWell", "BestCare"))
```

This example ordered the levels in the sequence of 1, 2 and 3 because the levels were listed in that order defined by the vector `1:3`. Other vectors could have been entered, such as `c(3,1,2)`, to specify a different order. Regardless of the specified order of the levels, the ordering of the labels must match the same ordering so that each label matches its corresponding level.

The labels applied in this example attached to integers. The `labels` parameter can also apply to variables of type of `character`. In that situation, display the vector of original character values as levels with a vector of another set of labels. For example, for the categorical variable *Gender*, display on the output a data value of `"M"` with the value label `"Male"`.

Create factors for many variables. Consider the survey data of the Mach IV items responded to on a 6-pt Likert scale from Strongly Disagree (0) to Strongly Agree (5). Obtain some analyses of these data (e.g., factor analysis) using the original numerical version of the responses as read from the data file. Perform other analyses such as bar charts with meaningful labels, that is, plot the factor version of the variables. In this situation, leave the original numerical scored items unchanged with their factor transformations defined as new variables.

*read Mach IV survey
data*, Section 1.2.4,
p. 17

The `lessR factors()` simplifies the creation of factor variables, particularly for the conversion of a set of multiple variables, either a sequence of consecutive variables in the data frame or a vector of individually specified variables. The first argument of the function call is the `x` parameter, which specifies the names of one or more variables to convert to factors. Here specify the variable range as m01:m20, which includes all contiguous variables in the data frame from m01 through m20. With so many categories it is more convenient to separately define the categories as a single character vector of the six labels, here called LikertCats.

R Input *Create a set of factor variables*

data: d <- Read("http://lessRstats.com/data/Mach4.csv")

```
LikertCats <- c("Strongly Disagree", "Disagree", "Slightly Disagree",
  "Slightly Agree", "Agree", "Strongly Agree")
d <- factors(m01:m20, levels=0:5, labels=LikertCats)
```

The previous function call replaces all 20 integer-scored variables with 20 factors labeled according to the LikertCats labels.

By default, the `factors()` parameter `new` is set to `FALSE`, which means that `factors()` replaces the original integer coded variables with the new factor variables. Set `new` to `TRUE` to leave the original variables unmodified and instead append the newly created factor variables to the original data frame. The newly created variables are by default named according to the original variables with the added suffix `.f`. Moreover, if variable labels are present for the original variables, then they can also be automatically moved to describe the parallel newly created factor variables.

Add levels

Sometimes not all possible responses for a categorical variable occur for one or more variables. The resulting visualization should usually include an analysis of potential responses, including those for categories with zero response. To do so, R must be made aware of potential data values that do not exist in the data.

An example is a self-report survey administered with the responses to items on Likert scales, such as Strongly Disagree, Disagree, Neutral, Agree, and Strongly Agree. Numerically code these responses such as with the integers from 1 (Strongly Disagree) to 5 (Strongly Agree). Visualizations of these scale responses, such as a bar chart, visualize the frequency of response to each of the alternatives. If no one responds to an item with Strongly Disagree, then the data for that item only consists of the integers from 2 to 5. Visualizations of the frequency of integer responses to that item do not include a 1 because without additional information, R only provides the frequencies for the data values that do occur.

To address this issue, use `factors()` to define a set of factors with the same set of possible responses for all the variables, such as items with the same set of Likert scale responses. These factors define response alternatives for potential responses that do not exist in the data. Visualization of responses for all the items are then of the same response categories, with 0 frequencies reported for response alternatives with no responses.

1.2.7   Save the Data Frame

The data have been read, cleaned, organized, and transformed, such as converting categorical variables to factors. Typically accomplish this work from a file of R instructions separate from the primary analyses that include the data visualizations. After the data preparation is complete, save the resulting data frame so to directly access from subsequent analyses without repeating all of the data preparation steps.

The most efficient file format for which to save the data frame is native R, a literal image of the data frame in an external file. The lessR function Write() accomplishes this save with format set to "R", abbreviated as wrt_r(). The written data frame, with filetype .rda, can then be easily read back into R in a later session with a matching lessR call to Read(), illustrated for a created data file named *attitudes*. The data frame written to that file is named the default name, *d*. To reference a data frame with another name, add the data parameter to the Write() function call.

Read(), Section 1.2.4, p. 12

R Input *Write/Read a native R data file*

```
lessR: Write("attitudes", format="R")
...
lessR: Read("attitudes.rda")
```

Another useful format for which to write the data is Excel[4]. Follow the same procedure as writing the data frame to native R except here set the format to "Excel", or use the abbreviation wrt_x(). If the data are to be read back into R from Excel, Write() can be paired with Read(). For efficiency, however, the generally recommended procedure uses the native R format if the written data are to be read back into R.

R Input *Write/Read an Excel file*

```
lessR: Write("Mach4", format="Excel")
...
lessR: Read("Mach4.xlsx")
```

The data written to the Excel file are formatted as shown in Figure 1.7.

	A	B	C	D	E	F	G	H	I	J	K	L	M	N	O	P	Q	R	S	T	U
1	Gender	m01	m02	m03	m04	m05	m06	m07	m08	m09	m10	m11	m12	m13	m14	m15	m16	m17	m18	m19	m20
2	0	0	4	1	5	0	5	4	1	5	4	0	0	0	0	4	0	1	3	2	4
3	0	0	1	4	4	0	3	3	0	4	4	0	1	1	1	2	4	4	3	1	0
4	1	2	1	0	5	4	4	0	5	3	4	1	4	0	0	2	0	2	4	0	1
5	1	0	5	2	4	0	4	4	4	5	2	0	0	0	1	1	1	5	4	4	0
6	0	2	2	3	3	2	3	2	3	1	4	1	3	1	2	2	2	3	3	2	1
7	1	1	3	3	5	1	3	3	1	5	3	0	0	0	3	2	5	2	3	1	0

Figure 1.7: First six rows of the Mach4 data and variable names from an Excel data table written from lessR Write().

With the R basics now presented, we are ready to create data visualizations.

---

[4]The lessR function Write() relies upon functions in Alexander Walker's openxlsx package to provide the Excel interface.

Chapter 2

Visualization Quick Start

Data analysis typically begins with the pursuit of answers to three fundamental questions. Seeking answers to these questions motivates three corresponding categories of visualizations.

1. *Question*: How are the data values for a variable patterned?

   Answer: Distribution of the values of a variable
   Visualization, categorical variable: Bar chart, dot plot, bubble plot
   Visualization, continuous variable: Violin plot, boxplot, scatterplot,
   density plot, histogram, frequency polygon

2. *Question*: Are two or more variables related?

   Answer: Joint distribution of the values of the variables
   Visualization, categorical variables: Stacked bar chart, bubble plot, mosaic plot
   Visualization, continuous variables: Scatterplot

3. *Question*: How do the values of one or more variables change over time?

   Answer: Distribution of the values of the variables over time
   Visualization, continuous variable: Run chart, time series chart

This chapter introduces visualizations that answer each of these fundamental questions. More detail and more answers follow in subsequent chapters.

2.1   Visualization Systems

This book includes extensive discussions and visualizations from what has become the standard source of R visualizations, ggplot2, as well as equivalent visualizations obtained from the more straightforward, though not as customizable, lessR package.

2.1.1   Relative Advantages of ggplot2 and lessR

Hadley Wickham's ggplot2 package provides the visualization functions most used by data scientists who use R. It is the most downloaded R contributed package, an impressive accomplishment with well over 10,000 available R packages on the CRAN servers. With ggplot2 Hadley Wickham implemented the conceptual work of Leland Wilkinson (2005), which articulates a general grammar of graphics (the gg in ggplot2).

All visualization software creates a visualization in successive layers, step-by-step, one feature at a time. For ggplot2, the analyst explicitly controls the implementation of this general grammar with a separate function call to implement each layer. The result is a toolkit, a set of building blocks available to the user that provides the basis for an extraordinarily wide range of both standard and highly customized visualizations.

lessR theme: Minimum computer input should provide maximumly useful computer output.

lessR and ggplot2 implement complementary approaches to data visualization. By design, lessR flips some of the perspectives of ggplot2, and, indeed, the perspective of R in general. The result is a different user experience to obtain the same quality of output for a wide variety of commonly employed visualizations. The primary distinctions include the following.

ggplot2 theme: Implement a wide variety of visualization functions for maximum flexibility and creativity.

1. Flexibility vs. minimal code. ggplot2 is a high-level graphics programming language. "High-level" means that some visualizations can be obtained with as few as three function calls on one line of code, but more complex, customized visualizations can require tens of lines of ggplot2 function calls and parameter specifications, with access to many functions. The core motivation underlying lessR is for the user to enter the smallest amount of code possible to obtain the most useful and comprehensive output. lessR provides a broad range of visualizations from only three core visualization functions: BarChart(), Histogram(), and Plot(), plus getColors() to generate color palettes and factors() to easily represent categorical variables as R factor variables for their proper representation in a visualization.

Locate information in the help files for these three core lessR visualization functions to customize a wide variety of visualizations. The list of parameters for each of the lessR visualization functions is long, but to simplify access, each corresponding help file groups related parameters together. The downside is less flexibility for which ggplot2 excels, of particular consequence if lessR does not provide the desired visualization that ggplot2 can construct with the proper coding.

color palettes, Section 10.2, p. 214

factors, Section 1.2.6, p. 19

2. Bottom-up vs. top-down. The traditional approach to constructing visualizations, such as employed by ggplot2, is what may be called bottom-up: Construct the visualization from its most basic form. Add enhancements by invoking additional options. For ggplot2, these additions consist of calling other functions to provide additional layers of the visualization, or customizing a given layer. The flipped perspective of lessR follows a top-down approach: The form of the visualization perceived as the most desired is provided by default, or as a simple option that provides many enhancements simultaneously. To change the default typically assigns parameter values to remove unwanted components.

The traditional bottom-up approach requires the user to focus on constructing the visualization at a higher level of abstraction. With this approach, explicitly code for each aspect of the visualization. The advantage is much potential for customization. The disadvantage is that the bottom-up approach requires more work to obtain the same result if available from an equivalent top-down visualizations.

3. Visualization only vs. plus statistics. lessR is not a visualization package per se, but a data analysis system for a core set of analyses, which includes visualizations as well as corresponding statistical analysis. For example, to conduct a full regression analysis in R – which includes collinearity analysis, outlier/influence analysis, and prediction intervals, as well as several visualizations – requires up to 20 different R functions. In contrast, the lessR function Regression() provides the entire analysis, including multiple visualizations, with one function call that follows standard R syntax.

lessR also includes other features. The lessR functions trap many R errors so that, in place of the usually cryptic R error message, a more extensive instruction is provided as to the nature of the error and its remedy. A default data table (frame) name, *d*, implies that only the function name and variable name(s) are required to generate the analysis. Again, however, `ggplot2` remains the ultimate for user flexibility for generating visualizations.

2.1.2   Grayscale

grayscale: Display only in black and white with intermediate shades of gray.

color palettes, Section 10.2, p. 214

Many publication sources such as printed academic journals, and, for example, this book, use grayscale illustrations. Modern visualization systems provide for both color and grayscale visualizations.

`lessR` uses color for visualizations as its default theme. Use the `lessR` function `style()` to obtain grayscale, which then applies to all subsequent visualizations.

> *grayscale theme*
>
> *lessR*: `style("gray")`

`ggplot2` visualizations tend to default to grayscale except for colors to distinguish among multiple levels of a categorical variable. Where color does appear, the `ggplot2` instructions throughout this book indicate the needed modifications to obtain grayscale visualizations. There are several variations. One example adds the following to the `ggplot2` function calls to create a bar chart.

> *grayscale theme*
>
> *ggplot2*: `scale_fill_grey(start=.6, end=.3)`

Next, begin creating visualizations.

2.2   Distribution of a Categorical Variable

bar chart: Associate a numerical value proportional to the height of a bar, for each value of a categorical variable.

bar chart, Section 3.1.1, p. 47

read employee data, Section 1.2.4, p. 12

lessR grayscale, Chapter 10.3, p. 223
default: Given values of parameters made active unless the user overrides.

Associate a numerical value with each category (or level or value) of a categorical variable, to create, for example, a bar chart, one of the most prominent data visualizations. The *bar chart*, displays a bar of proportionate height for each category according to the numerical value associated with the category. One commonly plotted numerical variable is the Count, the tabulation of how many data values occur at each level of the categorical variable. Chapter 3 explores the different visualizations of this type in more detail.

2.2.1   Bar Chart of a Single Variable

Consider the employee data set, which contains variables such as *Salary*, an employee's annual salary, and *Dept*, the department in which the employee works. How many employees work in each department? To answer, first read the data values from an external file into the *d* data frame. Then generate a bar chart. Figure 2.1 shows the default `lessR` grayscale bar chart and the default `ggplot2` bar chart.

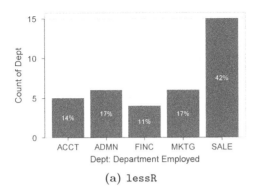

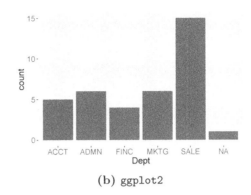

(a) `lessR`   (b) `ggplot2`

Figure 2.1: Default one-variable bar charts.

Default means that a visualization function sets the visual aesthetics such as the color of the bars, though all such aesthetics are customizable. By default, `ggplot2` plots the number of `NA` values, the missing data values, those Not Available. In contrast, `lessR` lists the number of missing values as part of the text output at the R console, shown in Listing 2.1.

ggplot2 color bar chart, Section 10.2.1, p. 217

The function calls that generate these default `ggplot2` and `lessR` bar charts follow. First, invoke the appropriate `library()` calls to access these functions from their corresponding packages. Reading the variable labels is optional, and only applies to `lessR`, but yields more complete labeling of the variable axis. The `lessR Read()` function separately reads both the data and the variable labels into the respective data frames named *d* and *l*.

variable labels, Section 1.2.5, p. 18

library(), function Section 1.1.2, p. 6

> **R Input** *Default bar chart of Count for variable* Dept *in* d *data frame*
> *data*: `d <- Read("http://lessRstats.com/data/employee.csv")`
> *labels*: `l <- Read("http://lessRstats.com/data/employee_lbl.csv"),`
>   `var_labels=TRUE`
> ──
> *lessR*: `BarChart(Dept)`
> *ggplot2*: `ggplot(d, aes(Dept)) + geom_bar()`

lessR

The `lessR` function names follow the *CamelCase* style, or medial capitals, which capitalizes each of one or more words with no intervening spaces. The `lessR` function `BarChart()` generates a bar chart. Alternatively, specify each `lessR` function with a two- or three-letter lowercase abbreviation, `bc()` for `BarChart()`.

CamelCase: A name consists of one or more words, each capitalized with no intervening spaces.

`BarChart()` also displays the corresponding descriptive and inferential analysis of the frequency distribution, shown in Listing 2.1 for this example.

The call to `BarChart()` in this example results from several abbreviations and conventions shown in Table 2.1. One convention relies upon the `lessR` default data frame (table) name, *d*. Or, explicitly reference a data frame with the `data` parameter in a `lessR` function call.

BarChart() function, `lessR`: Generate a bar chart and related statistics.

```
--- Dept: Department Employed ---

Missing Values of Dept: 1

                 ACCT    ADMN    FINC    MKTG    SALE     Total
Frequencies:        5       6       4       6      15        36
Proportions:    0.139   0.167   0.111   0.167   0.417     1.000

Chi-squared test of null hypothesis of equal probabilities
  Chisq = 10.944, df = 4, p-value = 0.027
```

Listing 2.1: BarChart() statistical analysis.

Function Call	Explanation
BarChart(x=Dept, data=d)	Parameter and data table both explicitly specified
BarChart(Dept, data=d)	Unlabeled value in first position is the x-variable
BarChart(Dept)	Default lessR data table is d
bc(Dept)	Abbreviation for BarChart() is bc()

Table 2.1: Four equivalent lessR function calls to generate the same bar chart and analysis.

If there is no y variable specified, as in this example, BarChart() computes the values of y from the tabulation of the counts of the values at each level of x.

ggplot2

layer: A plotted set of geometric objects that correspond to specified data values.

ggplot2 explicitly plots each set of visual objects as a distinct *layer*. All visualization packages plot the various components of a visualization one at a time. The compelling feature of ggplot2 is the control provided to the user for the construction of these layers. For example, each layer may be constructed from potentially different data, with different geometric objects plotted, a blending of multiple visualizations into a single composite. The analyst gains much flexibility with the freedom to add meaningful specifications in separate layers, instead of constrained to the specific structure provided by generic plotting functions.

ggplot2 function, **ggplot()**: Create a ggplot2 object to plot, perhaps identifying the data frame and associated variables to plot.

To create a ggplot2 visualization, first call the ggplot() function. Include a specification of the data frame that contains the variables for analysis. Next, specify the data values of one or more variables to plot with the embedded aes() function. Then specify the geometric object to plot, such as points or bars.

The current example in Figure 2.1b plots just one layer, the bars of a bar chart. If the same data frame and associated variables apply to all layers in the plot, then generally specify the data frame and its constituent variables within the call to ggplot(). Figure 2.2 illustrates this pattern that creates the bar chart in Figure 2.1.

visual aesthetics, Section 1.1.1, p. 2

A visualization consists of visual aesthetics such as coordinates on the x and y-axes, border color and fill color of plotted objects, and the shape of the plotted objects such as a dashed line, or a diamond for a plotted point. The grammar of graphics explicitly associates these aesthetics, what the viewer perceives, to the properties of the data values for one or more variables. This transformation of variables to visual aesthetics defines a *mapping*. At a minimum, the mapping associates the variables

```
    data frame  variable    object to plot
        —        ———      ——————————
    ggplot(d,  aes(Dept))  +  geom_bar()
```

Figure 2.2: `ggplot2` function calls to generate a bar chart from the tabulation of the count for each level of the categorical variable *Dept* in the *d* data frame.

plotted with the corresponding axes, and then applies default aesthetic values such as color and size if not explicitly stated.

As Figure 2.2 illustrates, one of the parameter values to `ggplot()` is another function, `aes()`. Specify visual aesthetics in `ggplot2` with the `aes()` function. The default first argument to the `aes()` function is the variable on the *x*-axis. A variable name as the first argument in the parameter list not preceded with an `x=` is implicitly defined as the variable mapped to the *x*-axis. Any second unnamed argument, if present, is the *y*-axis variable.

Typical geometric objects to plot include bars, lines, and points. Plot at least one geometric object in a layer that corresponds to some property of the data specified by the aesthetics. The `ggplot2` system refers to these geometric objects as `geoms`. Specify a `geom` by its own function, such as `geom_bar()` to plot the bars of a bar chart. Figure 2.2 includes all three functions that specify the resulting bar chart. Many other default values are also invoked.

The function calls in Figure 2.2 follow the useful R convention that the parameters do not need to be named if they follow the order of the parameters in the function definition. The first two `ggplot()` parameters are respectively named `data` and `mapping`. The `aes()` function maps, that is, transforms, the data values to the aesthetics as plotted to the specified geometric object, e.g., a gray bar. The first parameter for the `aes()` function is the variable *x*.

The function call that generates Figure 2.1 with all given parameter values named follows.

ggplot2: `ggplot(data=d, mapping=aes(x=Dept)) + geom_bar()`

For aesthetics unique to a single layer in a multi-layer plot, call the `aes()` function within that layer's `geom` function. For example, because there is only one layer in the Figure 2.1 plot, the following statement generates the same figure. Here the call to `ggplot()` specifies the default data table (frame) for all of the layers, *d*, but the call to `aes()` specifies the variable *Dept* only within the `geom_bar()` layer. Without a reference to a data frame in `geom_bar()`, the referenced variable, *Dept*, must be in the previously referenced data frame *d*.

ggplot2: `ggplot(d) + geom_bar(aes(Dept))`

In the following example, there is no default data table for the layers of this visualization because there is no `data` argument for `ggplot()`. Here specify both

---

mapping: Associate elements of one set, values of variables, with elements of another set, aesthetic properties of a visualization.

aes() function, **ggplot2**: Specify the visual aesthetics.

geom, ggplot2: A geometric object to plot.

parameter list, Figure 1.4, p. 15

Modification of individual ggplot2 styles, Section 10.4.1, p. 226

the variable plotted and the name of the data table in the specific `geom` function that defines the plotted object for the layer. The following line of code also generates the same Figure 2.1 because this visualization only contains a single layer.

```
ggplot2: ggplot() + geom_bar(aes(Dept), data=d))
```

To apply these specifications to subsequent layers, unless explicitly overridden by new specifications, include in the originating `ggplot()` function call both the variable(s) with the data values to plot, as well as the data frame that contains these variables. This flexibility provides for different data frames and variables specified as needed in different layers, a topic explored later.

2.2.2 Bar Charts of Multiple Variables

Multiple single bar charts

vectors, Section 1.2, p. 6

The `lessR` function `BarChart()` provides bar charts for as many variables as specified. To process more than a single variable, set the value of the x parameter, the first parameter in the parameter list, to a vector. Or, to produce a bar chart for every categorical variable in the data frame, with its corresponding frequency distribution, do not provide any value of x. The function call to generate a bar chart for each categorical variable in the d data frame, accepting the default values for all parameters, reduces to four characters:

```
lessR: bc()
```

factors, Section 1.2.6, p. 19

For one or more integer coded categorical variables not converted to factors, the `lessR` parameter `n.cat`, for the number of categories, informs R that these numerical variables are categorical. Specify the maximum number of unique integer codes that define a categorical variable. For example, the Mach IV data are coded on a 6-pt Likert scale with integers 0 through 5. The following creates bar charts for all 20 of the integer coded Mach IV items plus the 0-1 *Gender* variable in the data table.

read Mach IV survey data, Section 1.2.4, p. 17

```
lessR: bc(n.cat=6)
```

Stacked bar charts

stacked bar chart: Divide a single bar for a categorical variable into regions, one for each level.

For data sets with categorical variables that have different response scales, such as *Dept* and *Gender* from the Employee data set, each bar chart appears in a separate window by default. The `BarChart()` default changes, however, for a set of variables that share the same response scale, such as the 20 Mach IV items each assessed with a 6-pt Likert scale. Then `BarChart()` plots a single visualization as a set of *stacked bar charts*, one bar for each variable, here an individual item. Divide each bar into regions that correspond to the response categories.

Bubble Plot Frequency Matrix an alternative to the stacked bar chart, Figure 3.4, p. 50

The grayscale version of the stacked bar charts follows in Figure 2.3. The middle lighter colors diverge on each side to darker colors, here limited to grays. The default presents a yellow-brown progression on the left of each bar and a blue progression on the right.

divergent color scale, Section 10.2.3, p. 221

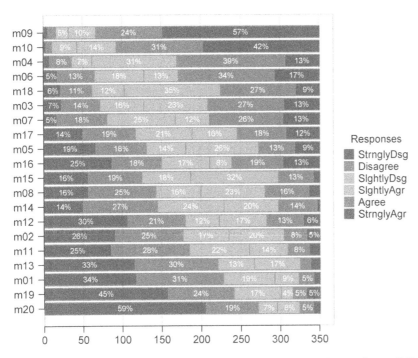

Figure 2.3: For each of the 20 Mach IV items, a stacked bar chart of 6-pt Likert scale responses sorted by mean level of agreement for 351 respondents.

`BarChart()` analyzes the responses to the categorical variables and then sets the parameter `one_plot` as `TRUE` or `FALSE` accordingly. Or, override the default as preferred. Even with different response scales, variables can be plotted on a single panel. Or, create separate visualizations for variables with the same response scale.

`lessR` visualizations automatically include generally desired features. For multiple stacked bar charts these features include horizontal bars, variables sorted by their mean responses, and a display of the numerical values on the bars. Explicitly remove added features not desired. In this example, set `horiz` to `FALSE`, `sort` to 0, and `values` to `"off"`.

Obtain the stacked bar charts in Figure 2.3 with a call to `BarChart()`. Replace a reference to a single variable with a vector of variables with the same response scale.

vector, Section 1.2, p. 6

> **R Input** *Visualization of stacked bar charts with the same response scale*
>
> *data*: `d <- Read("http://lessRstats.com/data/Mach4.csv")`
>
> ---
>
> *lessR*: `BarChart(m01:m20)`

variable labels, Section 1.2.5, p. 18

Output to the console includes the frequency distribution for each variable and its corresponding mean. If variable labels are included, such as the items on the Mach IV scale, `BarChart()` writes the items to the console in the order of presentation in the stacked bar charts.

Centered stacked bar charts

The `lessR BarChart()` plots the stacked bars against the cumulative sample size on the horizontal axis. The reported value for each response category is the cumulative sum of the number of responses for that category and all previous categories. Accordingly, each bar extends the full horizontal width of the horizontal axis, bounded by the total sample size, here $n = 351$.

sequential scale,
Section 10.2.2, p. 218

divergent scale,
Section 10.2.3, p. 221

Figure 2.4 demonstrates an alternate approach, which centers each bar at the neutral point, marked by a vertical line through all the bars, and uses a sequential scale instead of a divergent scale. Here the horizontal axis is bounded on both sides by the percentage of agreement and percentage of disagreement, respectively. In this example of a six-point response scale, beginning with 1 for Strongly Disagree and ending at 6 for Strongly Agree, the Disagreement responses – 1, 2, and 3 – are on the right side of the vertical line, and the Agreement responses – 4, 5, and 6 – are on the left side.

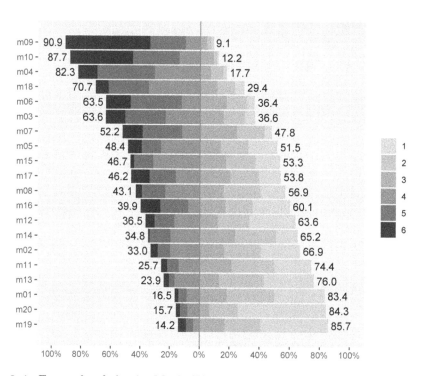

Figure 2.4: For each of the 20 Mach IV items, a centered stacked bar chart of 6-pt Likert scale responses sorted according to ascending order by sum of positive responses for 351 respondents.

plot_likert()
function, sjPlot:
Plot responses from
multiple Likert
response items on
the same panel as
centered stacked bars.

Extract() function,
Section 1.2.4, p. 13

The `plot_likert()` function from Daniel Lüdecke's (2018) package `sjPlot` created Figure 2.4. The package provides many plotting routines for a variety of statistical analyses. Input to `plot_likert()` the data values in the specific columns that correspond to the variables to plot. Here the data values for the 20 items of the Mach IV scale are contained in columns 2 through 22 of the input csv data file. Read the data into data frame *d*, then select the relevant columns with the base R `Extract()` function, indicated by the square brackets.

R Input *Visualization of centered stacked bar charts*

data: `d <- Read("http://lessRstats.com/data/Mach4.csv")`
      `d <- d[,2:21]`
lessR: `d <- Recode(m01:m20, old=0:5, new=1:6, data=d)`

---

sjPlot: `plot_likert(d, show.n=FALSE, geom.size=.9, sort.frq="pos.asc",`
        `values="sum.outside", reverse.colors=TRUE, geom.colors="gs")`

The 6-pt Likert scale responses are coded 0 through 5. Apparently, `plot_likert()` requires responses to begin at 1, so recode the responses to the new metric, here using the `lessR` function `Recode()`. To use this function, list the relevant variable names as a vector, then the **old** codes, and then the corresponding **new** codes. Specify the input data frame name *d* with the **data** parameter, which is, however, not required for the default name of *d*.

With the data read and prepared, invoke `plot_likert()` beginning with the name of the revised data frame, **d**. Additional modifications for this example begin with not showing the *n = 351* output for each line, by setting `show.n` to `FALSE`. Increase the bar heights a bit with `geom.size` set to 0.9. Sort in ascending order according to the sum of the Agree answers, which is a different sort criterion than used by `lessR`, so the sorted orders differ a bit. Show the sums for Agree and then Disagree responses with `values` set to `"sum.outside"`. Obtain grayscale color with `geom.colors` set to `"gs"`. Reverse the shading of the grayscale palette with `reverse.colors` set to `TRUE`.

2.3   Distribution of a Continuous Variable

The histogram displays the distribution of a continuous variable, such as *Age*, *Salary*, *MPG* or *Height*. Partition the underlying continuity that spans the range of data values into intervals generally of equal width called *bins*. Place each data value in its corresponding bin, then count the number of values in each bin. To construct the histogram, usually place the bins on the horizontal axis. Construct a bar over each bin to represent its corresponding frequency. Unlike a bar graph of a categorical variable, indicate the underlying continuity of the numerical scale with adjacent bins that share a common border.

bin: An interval of similar values of a continuous variable.

2.3.1   Default Histogram

The default grayscale `lessR` histogram, and a `ggplot2` histogram, appear in Figure 2.5.

The histogram of *Salary* shows that for these 37 employees the most frequently occurring salaries are from \$40,000 to \$60,000, with relatively few salaries over \$100,000.

histogram: Place each data value in its corresponding bin, represented by a bar with a height proportional to the frequency of its values.

workspace,
Section 1.2.1, p. 7

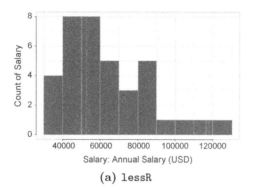

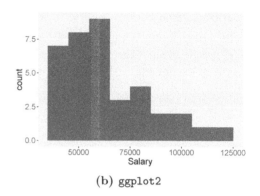

(a) `lessR`   (b) `ggplot2`

Figure 2.5: Default `lessR` histogram, and `ggplot2` histogram with set bin width.

Abbreviate the `lessR` histogram function `Histogram()` as `hs()`. Locate the relevant variable within the user workspace directly, or within a data frame, here with the default name of *d*. For `ggplot2`, plot the histogram layer with the `geom_histogram()` function that follows the same general form as illustrated in Figure 2.2.

> **R Input** *Default histogram:* Salary *from* d *data frame*
> *data*: d <- Read("http://lessRstats.com/data/employee.csv")
>
> *lessR*: Histogram(Salary)
> *ggplot2*: ggplot(d, aes(Salary)) + geom_histogram(binwidth=10000)

In practice, default algorithms for bin width generally provide an acceptable, but not necessarily optimal, bin width. After plotting the initial histogram, explore various bin width alternatives. To encourage this exploration, the `ggplot2` histogram function does not provide a reasonable default bin width. To specify the bin width, set the `binwidth` parameter in the `geom_histogram()` call. The equivalent `lessR` parameter is `bin_width`.

The `lessR` `Histogram()` output also includes a frequency distribution table of the bins, bin midpoints, frequencies, relative frequencies or proportions, and cumulative relative frequencies, as well as summary statistics, and an outlier analysis.

VBS Plot:
Integrated violin, box
and scatterplot for
the display of the
distribution of a
continuous variable.

2.3.2   Beyond the Histogram

Modern computer technology provides for more effective visualizations than the 19th-century histogram, such as with the VBS plot for the display of the distribution of a continuous variable (Gerbing, 2020). The VBS plot integrates three different but complementary plots into a single visualization: A violin plot (Hintze & Nelson, 1998), a box plot (Tukey, 1977), and a 1-variable scatterplot. The plotting function then automatically tunes the completed visualization in terms of the sample size and other characteristics of the distribution. Find a more extensive discussion of all three subplots and of the VBS plot in Chapter 4 beyond the following introduction.

VBS plot,
Section 4.5, p. 94

Figure 2.6 illustrates the `lessR` and `ggplot2` VBS plots for a small data set of 37 annual salaries.

```
--- Salary: Annual Salary (USD) ---

    n    miss         mean          sd         min         mdn         max
   37       0     63795.56    21799.53    36124.97    59547.60   124419.23


(Box plot) Outliers: 1

Small       Large
-----       -----
            124419.2



Bin Width: 10000
Number of Bins: 10

              Bin   Midpnt   Count    Prop  Cumul.c  Cumul.p
-------------------------------------------------------------
  30000 >   40000    35000       4    0.11        4     0.11
  40000 >   50000    45000       8    0.22       12     0.32
...
 110000 >  120000   115000       1    0.03       36     0.97
 120000 >  130000   125000       1    0.03       37     1.00
```

Listing 2.2: Statistical analysis provided by the `lessR` function `Histogram()`.

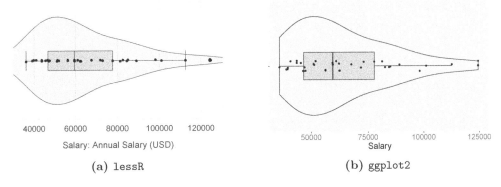

(a) `lessR`   (b) `ggplot2`

Figure 2.6: Integrated Violin/Box/Scatterplot or VBS plot.

The VBS plot of *Salary* shows that for these 37 employees, the most frequently occurring salaries range from \$40,000 to \$60,000, with relatively few salaries over \$100,000. The boxplot marks the highest value of *Salary*, over \$120,000, as an outlier.

outliers, one-variable, Section 4.13, p. 92

outliers, two-variables, Section 5.6, p. 109

The VBS plot demonstrates a primary goal of `lessR`: Conceive of what is most useful, and then provide the relevant visualization with a simple function call given appropriate defaults. The `lessR` function `Plot()` generates the VBS plot from the values of a single continuous variable. `Plot()` by default adds each component of the plot – violin plot, box plot, and scatterplot – as a separate layer, and automatically adjusts characteristics of the plot according to the sample size and distribution type. In contrast, to use `ggplot2` to construct the VBS plot, explicitly add each of the three components as separate layers, each with its own `geom` and then run successive visualizations manually tuning the parameters such as bandwidth and point size to obtain the desired result.

ggplot2 VBS code detail, Figure 4.16, p. 95

> **R Input** *VBS Plot of* Salary *from* d *data frame*
>
> *data*: d <- Read("http://lessRstats.com/data/employee.csv")
>
> ---
>
> *lessR*: Plot(Salary)
>
> *ggplot2*: ggplot(d, aes(x="", y=Salary)) +
>   geom_violin(fill="gray90", bw=9500, alpha=.3) +
>   geom_boxplot(fill="gray75", outlier.color="black", width=0.25) +
>   geom_jitter(shape=16, position=position_jitter(0.05)) +
>   theme(axis.title.y=element_blank()) +
>   coord_flip()

To re-iterate a basic visualization theme: lessR provides the simplicity to obtain a pre-programmed result and ggplot2 provides the flexibility, at the cost of additional complexity, to create what can be conceptualized.

2.4 Relation between Two Variables

scatterplot: For each observation, the values of n variables plotted as a point in an n-dimensional coordinate system.

Represent data values as plotted points with the *scatterplot*, a 19th-century invention by Francis Galton (Friendly & Denis, 2005). The two-dimensional scatterplot *relates* two continuous variables by their joint data values, with an axis for each variable. Each point on a two-dimensional scatterplot represents the values of the two variables for a single observation. Plot the respective values for the variables as the coordinates of the two axes.

relationship: As the values of one variable increase, the values of the other variable tend to either systematically increase or decrease.

2.4.1 Basic Scatterplot

The scatterplots in Figure 2.7 demonstrate a strong, positive relationship between *Years* employed at the company with *Salary*.

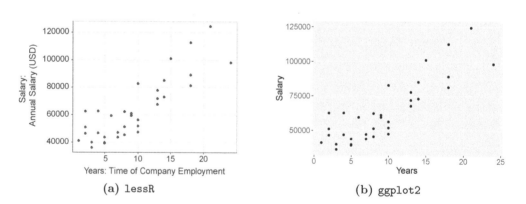

(a) lessR   (b) ggplot2

Figure 2.7: Default scatterplots.

scatterplots,
Chapter 5, p. 103

The lessR function Plot(), abbreviated sp(), creates a scatterplot, broadly defined to include a wide variety of visualizations with plotted points, including any combination of continuous and categorical variables. To create a two-variable scatterplot, such as for the two continuous variables in this example, specify both an x and a y variable, the first two arguments of the function Plot().

For ggplot2, define the data frame and variables in the call to ggplot(). Specify the x and y variables with the aes() function, the first two arguments of the function. The function geom_point() specifies to plot points, the single layer in this visualization.

> **R Input** *Default scatterplot of* Years *and* Salary
>
> *data:* d <- Read("http://lessRstats.com/data/employee.csv")
>
> ---
>
> *lessR:* Plot(Years, Salary)
> *ggplot2:* ggplot(d, aes(Years, Salary)) + geom_point()

Plot() also provides an analysis of the correlation coefficient. In addition to the estimated coefficient value, the analysis includes the hypothesis test of zero for the population correlation, and the accompanying confidence interval, as shown in Listing 2.3.

```
Number of paired values with neither missing, n = 36

Sample Correlation of Years and Salary: r = 0.852

Alternative Hypothesis: True correlation is not equal to 0
  t-value: 9.501,  df: 34,  p-value: 0.000

95% Confidence Interval of Population Correlation
  Lower Bound: 0.727      Upper Bound: 0.923
```

Listing 2.3: Plot() statistical output for the analysis of the correlation.

Computer technology provides enhancements to the scatterplot not possible with the 19th-century technology, the next topic.

2.4.2 Enhanced Scatterplot

Standard errors around best fit line, Chapter 3, p. 45

Enhance the scatterplot with several features. The 95% confidence ellipse from the bivariate normal distribution helps to visualize the relationship. The best-fit least squares line plotted with the corresponding 95% confidence bands summarizes the (linear) relationship. The identification of potential outliers with a re-computation of the best-fit line without the outliers present calibrates their influence on the estimated relationship. Delineating the four quadrants of the plot further assists in visualizing the nature of the underlying relationship.

Figure 2.8 illustrates all of these features for the lessR visualization and all except the outlier analysis for ggplot2. The corresponding ggplot2 visualization does not identify the outliers, nor present the alternative best-fit line without the outliers. The flexibility and freedom to customize ggplot2 imply that it could also create the same visualization in Figure 2.8a. However, to do so would require many more lines of code. First, calculate Mahalanobis distance of each point from the center to identify outliers, then delete the outliers, then the re-calculate and display the regression line for the data without the outliers.

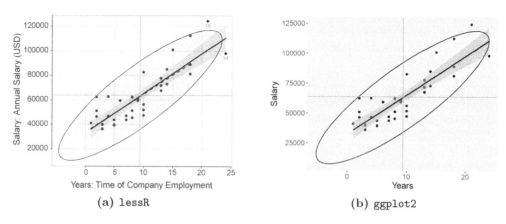

(a) <code>lessR</code>   (b) <code>ggplot2</code>

Figure 2.8: Enhanced scatterplots.

To create the enhanced `lessR` visualization, set a single parameter, `enhance`, to `TRUE`. For `ggplot2`, a distinct function call generates each layer of the resulting visualization in Figure 2.8b. If `lessR` has a parameter option for what is desired, then simpler to use, but for the many possibilities not pre-programmed into `lessR` functions, `ggplot2` provides the ultimate toolkit for generating customized visualizations.

R Input *Enhanced scatterplot:* Years *and* Salary

data: d <- Read("http://lessRstats.com/data/employee.csv")

---

lessR: Plot(Years, Salary, enhance=TRUE)

ggplot2: ggplot(d, aes(Years, Salary)) + geom_point() +
  geom_smooth(method=lm, color="black") +
  stat_ellipse(type="norm") +
  geom_vline(aes(xintercept=mean(Years, na.rm=TRUE)), color="gray70") +
  geom_hline(aes(yintercept=mean(Salary), na.rm=TRUE), color="gray70")

For `ggplot2`, obtain the best-fit least-squares line and associated default 95% confidence interval with `geom_smooth()`. By default, the line is plotted in blue, so obtain grayscale by specifying a color such as `"black"`. Obtain the 95% confidence ellipse with `stat_ellipse()`. The `type="norm"` argument specifies the assumption of multivariate normality. Plot the vertical and horizontal lines with the `geom_vline()` and `geom_hline()` functions. The respective `xintercept` and `yintercept` arguments indicate where to place the lines on the corresponding axes. The `na.rm` parameters instruct R to calculate the mean in the presence of missing data. Otherwise, the computation fails in the presence of missing data.

2.5   Distribution of Values over Time

time series: Data values distributed over regular time intervals, and the corresponding visualization.

A *time series* displays time-oriented data ordered across the labeled corresponding time periods. Plot the time values on one axis and the values of the variable, one for each time period, on the other axis.

2.5.1 Time Series

To illustrate, consider the adjusted stock price obtained from `finance.yahoo.com` of three companies from 1980 though January of 2019: Apple, IBM, and Intel. Figure 2.9 shows that the Excel data file consists of three columns, for monthly *date*, *Company*, and corresponding *Price* per share of company stock. Consider only the time series for Apple in Figure 2.10, which shows the general dramatic rise in Apple's market value following the introduction of the iPhone in 2008.

1	date	Company	Price
2	12/1/1980	Apple	0.027
3	1/1/1981	Apple	0.023
4	2/1/1981	Apple	0.021
	...		
460	12/1/1980	IBM	2.051
461	1/1/1981	IBM	1.945
462	2/1/1981	IBM	1.941
463	3/1/1981	IBM	1.910
	...		
918	12/1/1980	Intel	0.212
919	1/1/1981	Intel	0.196
920	2/1/1981	Intel	0.185
921	3/1/1981	Intel	0.191
	...		

Figure 2.9: Time series data.

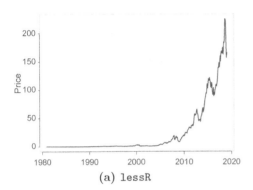

(a) `lessR`

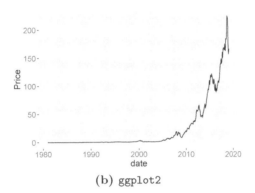

(b) `ggplot2`

Figure 2.10: Default time series plots.

To activate a default time series visualization, format the x variable that lists the dates as a date variable. If the data file read into R is an Excel file with the date variable formatted as an Excel date, then that formatting carries over to the R data frame. Otherwise, when reading an Excel file without the date variable formatted as a date, or reading a `csv` file, explicitly convert to a date variable.

date formats,
Section 7.2.4, p. 168

Select only the data for Apple. The data table consists of equal numbers of rows of data for each of the three companies, but the plotted time series in Figure 2.10 applies only to Apple's share price. To limit the visualization to Apple, select only the rows of data for Apple. For `lessR` visualization functions such as `Plot()`, set the `rows` parameter so that the value of Company equals `"Apple"`. R uses the double equal sign, `==`, to indicate "is equal to" for a logical comparison.

is equal operator, `==`, `TRUE` if equal expressions on both sides of the operator

```
rows=(Company == "Apple")
```

For `ggplot2`, select the specified data rows *before* invoking the `ggplot()` function, such as using the base R `subset()` function, or the tidyverse function `filter()` from the `dplyr` package, both of which have the identical syntax except for the function name. To invoke one of these functions, here use Stefan Bache's *pipe operator* from the `magrittr` package (Bache & Wickham, 2014), `%>%`, pronounced "then", that relates two functions. The operator takes the output of the function listed before it and, by default, inputs that output into the first parameter of the

pipe operator, `%>%`: Transfer output from one function as input to a subsequent function.

function that follows it. The separately available `magrittr` package for convenience also resides in the tidyverse.

Consider the expression without the pipe operator,

```
filter(d, Company == "Apple")
```

The first parameter value of the `filter()` function, *d*, is the value of the `data` parameter, which could also be written as `data=d`. Rewrite using `%>%` as,

```
d %>% filter(Company == "Apple")
```

With `%>%`, no longer specify the `data` parameter. The expression that precedes the `%>%`, *d*, indicates the information to input into the `filter()` function, specifically the information input into its first parameter value, `data`.

In general, `%>%` pipes the output from the first function to the parameter value of the second function specified with the value of the punctuation mark of a period. To not rely upon the default of inputting the information to the first parameter of the receiving function, rewrite the previous expression to explicitly include the parameter that receives the piped output from the previous function:

```
d %>% filter(data=., Company == "Apple")
```

By explicitly including the parameter to which to direct the input, that parameter could be located anywhere in the list of parameters in the function call, not just the first position.

Next pipe the output of the `filter()` function to the `ggplot()` function, as illustrated in the following code, which references a date variable already formatted as a date within the Excel data file.

> **R Input** *Default time series: monthly Apple share price*
> *data*: `d <- Read("http://lessRstats.com/data/PPStechLong.xlsx")`
> ──
> *lessR*: `Plot(date, Price, rows=(Company == "Apple"))`
> *ggplot2*: `d %>% filter(Company == "Apple") %>%`
>   `ggplot(aes(date, Price)) + geom_line()`

When `lessR` detects that the variable to plot is one of the R date types, by default it plots points with a `size` of 0, and adds the lines that connect what would be the plotted points. Explicitly plot a line in `ggplot2` with `geom_line()`. To plot both the points at the time values and the connecting line segments, in `lessR` specify a `size` larger than zero. For `ggplot2`, to obtain both points and lines, add the `geom_point()` function as a separate layer to the `ggplot()` specification.

panel: A single plot with its own axes, perhaps in the context of a family of related plots.

2.5.2 Multiple Time Series

Plot multiple time-series in the same visualization across a distribution of values over time across different levels of a second, categorical variable. Here, compare share price of each company over time. Obtain a time series for Apple, IBM, and Intel on the same visualization in one of two ways. For one option, place all three time series plots on the same coordinate axes, the same panel. Or, separate the plots into separate panels so that the visualization consists of multiple panels with each time series plotted with its own axes.

Single plot with a different result for each group

First consider overlapping plots on the same panel, as shown in Figure 2.11.

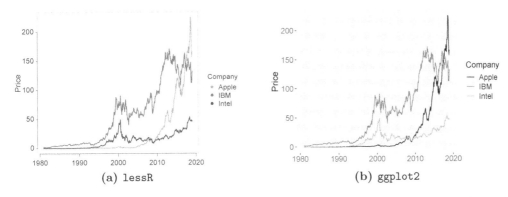

(a) `lessR`   (b) `ggplot2`

Figure 2.11: Time series plots for Apple, IBM and Intel on the same panel.

`lessR` uses the `by` parameter to specify a categorical variable to define separate subsets of the data to plot separately on the same panel. For the `ggplot2`, associate the categorical variable of interest, Company, with a visual aesthetic. For both visualization systems the aesthetic `color` refers to the color of an object as viewed from the outside, the edges of filled regions or lines by themselves.

R Input *Default time series: monthly Apple share price*

data: `d <- Read("http://lessRstats.com/data/PPStechLong.xlsx")`

---

lessR: `Plot(date, Price, by=Company)`
ggplot2: `ggplot(d, aes(date, Price, color=Company)) + geom_line()`

Each time series in a `ggplot2` multiple time-series plots with a different color. To specify grayscale lines, add a + to the end of the `ggplot2` instruction and then reference the `scale_color_gray()` function, here with custom **start** and **end** values.

grayscale,
Section 2.1.2, p. 28

customizations,
Chapter 10, p. 209

Added instruction to display the ggplot2 time series in grayscale

---

ggplot2: `scale_color_gray(start=0, end=.6)`

As in all the examples, to obtain grayscale with `lessR` invoke `style("gray")` before any instructions for visualizations.

Multiple panel plot

Trellis graphics: A
rectangular grid of
plots of the same
variables for each
level of one or more
categorical variables.

William Cleveland (1993) designed and implemented *Trellis Graphics* in the R-precursor, S. Deepayan Sarkar subsequently applied these visualizations in his (2008) `lattice` graphics package, which `lessR` relies upon for Trellis graphics. These visualizations provide a separate plot, called a *panel*, for each subset of data for each level of a categorical variable. Usually, the axes of the different panels share the same scale.

Trellis plot, a
formatted time series
version, Figure 7.7,
p. 167

Figure 2.12 illustrates a Trellis graphics multiple time series with `lessR` and `ggplot2`, a separate time series panel for the share price of Apple, IBM and Intel.

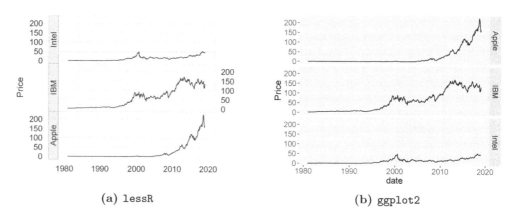

(a) `lessR`   (b) `ggplot2`

Figure 2.12: Default Trellis graphics time series plots, one panel each for Apple, IBM, and Intel.

The `lessR` functions implicitly activate Trellis graphics by a direct call to the corresponding lattice function. Specify the first conditioning variable with the `by1` parameter and the second conditioning variable with `by2`. To specify a single column of panels set `n.col` to 1. Or, use `n.row` to specify the number of rows.

`ggplot2` implements its version of Trellis graphics with panels named `facets`. Call the `facet_grid()` function to create Trellis graphics. To specify a single column of panels the value passed to the function begins with the name of the conditioning variable passed as a parameter value to the `vars()` function, specified with `col`.

R Input *Trellis plots*
data: d <- Read("http://lessRstats.com/data/PPStechLong.xlsx")

---

lessR: Plot(date, Price, by1=Company)
ggplot2: ggplot(d, aes(date, Price)) + geom_line() +
                facet_grid(rows=vars(Company))

Replace `rows` with `cols` in the call to `facet_grid()` to place the panels by columns.

Chapter 7 explores enhancements to these time series visualizations, which includes a discussion of the type of variable called an R time series, more detail regarding R date types, and options to provide alternative time series plots.

Chapter 3

Visualize a Categorical Variable

bar chart from the tabulation of values for a categorical variable, Figure 2.1, p. 29

coordinate data table: Rectangular table that pairs each category with a numerical value, one row for each unique category.

3.1   Bars, Dots, and Bubbles

As discussed in Chapter 2, an essential set of visualizations, such as the bar chart, relates each level (or category) of a categorical variable, x, to a value of an associated numerical variable, y. For example, consider an analysis of the class standing of the students in a college course,

Class	Count
UnderGrad	28
PostGrad	3
Grad	9

Figure 3.1: Coordinate data table.

which has enrolled 28 undergraduates, 3 post-graduates not part of a graduate program, and 9 graduate students. How many students are in each category of class standing? To visualize the relation between x and y, proceed from the pairing of the categories and the numerical values shown in Figure 3.1. The coordinates of the bar chart are derived from this table of categories paired with numbers, so refer to this data table as the *coordinate data table*.

For a bar chart, the numerical coordinates define the height of the bars. Construct the coordinate data table from any pairing of categories with numbers. Typically the numbers are statistical summaries, such as a tabulation of the occurrence of values of each category, or a value such as the mean of some variable such as salary for each category. The data in Figure 3.1, for example, were tabulated from the class roll of registered students. In this situation, the coordinate data table is a summary table, what Excel refers to as a pivot table.

If the coordinate data table is not entered directly into the visualization function, then the original data values of measurements must be transformed into a summary table, such as shown in Figure 3.1. The transformation can occur separately from the visualization function, using some of the many data manipulation tools provided by base R, `lessR`, and the tidyverse. Or, the visualization function provides this transformation, either by default or indicated by a specified parameter value.

Two primary classes of transformations apply to the data values in the original data table of measurements. The first set requires only the categorical variable x. In this situation, obtain the value of y as the tabulation of the number of occurrences for each level (category) of x, such as in Figure 3.1. The second class of transformations provides the values of the variable y from the original data table of measurements to the visualization function. Then the function transforms the values of y with a statistical summary, such as the mean, for each level of x. The computed coordinate data table then pairs each level of x with its corresponding numerical value, y.

This section reviews the bar chart, and related visualizations of the relationship with dots and bubbles. Each example either enters the coordinate data table directly, or explicitly or implicitly transforms the original data table of measurements into the coordinate data table.

3.1.1 Horizontal Bar Chart of Counts

The base R bar chart function, `barplot()`, which `lessR` relies upon to create a bar chart, only operates directly on the supplied coordinate data table. So any needed data transformation occurs before calling `barplot()`. In contrast, for transformations without an entered y-variable, both `lessR` and `ggplot2` by default tabulate the counts for each category to define y. For `lessR`, this tabulation occurs by default when no y variable is specified in the call to `BarChart()`. For `ggplot2`, this tabulation follows when no y variable is specified for the geometric object `geom_bar()`.

The Quick Start chapter, Chapter 2, presents the default vertical bar chart based on the default tabulation of the frequency of values of a categorical variable when no numerical variable, y, is specified for the analysis. Here present a bar chart with horizontal bars from the same data, as shown in Figure 3.2.

Bar chart: Associate a numerical value, proportional to the height of a bar, with each value of a categorical variable.

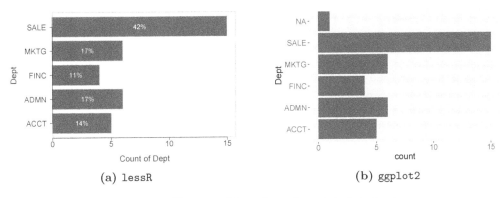

(a) `lessR`   (b) `ggplot2`

Figure 3.2: Horizontal bar chart from tabulated values.

To specify horizontal bars with `lessR`, set the `barplot()` parameter `horiz` to `TRUE`. For `ggplot2`, to obtain a horizontal bar chart add the `coord_flip()` function to the functions that generate a bar chart.

> **R Input** *Horizontal bar chart*
> *data*: `d <- Read("http://lessRstats.com/data/employee.csv")`
> ---
> *lessR*: `BarChart(Dept, horiz=TRUE)`
> *ggplot2*: `ggplot(d, aes(Dept)) + geom_bar() + coord_flip()`

coordinate data table,
Figure 3.1, p. 46

The tabulation of the count for each category occurs by default for each visualization system. The parameter `stat` explicitly specifies the statistical computation that provides the values of the y variable for the bar chart. For both `lessR` and `ggplot2`, the default value of `stat` is `"count"`. `ggplot2` sets the value of `stat` to `"count"` as part of the function call to `geom_bar()`. Both of the following function calls generate Figure 3.2, neither of which enters a value of y into the analysis, and neither of which requires the `stat` parameter to achieve the same visualization.

> *Generate horizontal bar chart from original data, specify the default values*
>
> *lessR*: BarChart(Dept, horiz=TRUE, stat="count")
>
> *ggplot2*: ggplot(d, aes(Dept)) + geom_bar(stat="count") + coord_flip()

3.1.2   Cleveland Dot Plot of Counts

scatterplot,
Section 2.4.1, p. 38

The widths of the bars for the bar chart in Figure 3.2 do not communicate any information. Accordingly, the Cleveland dot plot (Cleveland, 1993) replaces the bar chart with a scatterplot of the counts plotted against the levels of the categorical variable. Here enhance the visualization with each plotted point connected to the variable axis with a line segment, as illustrated in Figure 3.3.

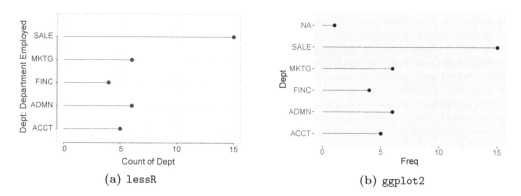

(a) `lessR`   (b) `ggplot2`

Figure 3.3: Cleveland dot plot of counts for *Dept*.

The `lessR` function `Plot()` constructs scatterplots. To inform `Plot()` to visualize counts calculated from the original data table of measurements as a Cleveland dot plot, do not specify a value of y. Then set the value of the `stat` parameter to either `"count"` for the frequencies, `"proportion"` for the relative frequencies, or `"%"` for the relative frequencies expressed as percentages. The `stat` parameter specifies the underlying transformation of the values of x, and for `Plot()`, indicates the form of the resultant visualization as a dot plot.

Instead of building the plot layer by layer, this `lessR` Cleveland dot plot follows the default of automatically creating additional layers that results in what is considered the most appealing or useful visualization. Instead of adding layers, the user is free to remove layers. Specify line segments in `Plot()` according to the `segments.y` and `segments.x` parameters. For most scatterplots, these parameters default to `FALSE`, but for the dot plot the relevant parameter, here `segments.y`, defaults to `TRUE`. Also, with so few points to plot, `Plot()` increases the usual default size of the plotted points for scatterplots.

%>% pipe operator,
Section 2.5.1, p. 41

To obtain the Cleveland dot plot with `ggplot2`, first generate the counts of the number of employees in each department, the frequency distribution for *Dept*. Here aggregate the data with the tidyverse `dplyr` functions `group_by()` to define the groups, and `summarise()` to obtain the statistical summary. Use the pipe operator, `%>%`, which, by default, transfers the output of the leading function to the first parameter of the following function. Name the newly created variable *Freq*.

Plot the `ggplot2` scatterplot with `geom_point()`. Specify the two names of the variables to plot either in `ggplot()`, or if applicable only to this layer, directly in the `geom_point()` layer. With only five points to plot, increase the size of the plotted points. Generate the line segments with `geom_segment()`. The input here is the starting point of a line segment, specified by the x and y parameters, and the corresponding end point, `xend` or `yend`. Reduce the default width of the segments in this example to 0.25.

R Input *Cleveland dot plot of Dept counts*

data: `d <- Read("http://lessRstats.com/data/employee.csv")`

---

lessR: `Plot(Dept, stat="count")`

ggplot2: `d %>% group_by(Dept) %>% summarise(Freq=n()) %>%`
`   ggplot() + coord_flip() +`
`     geom_point(aes(Dept, Freq), size=2.25) +`
`     geom_segment(aes(x=Dept, y=0, xend=Dept, yend=Freq), size=.25)`

Enter the parameter values of `geom_segment()` from the perspective of a vertical orientation, which is why the beginning y coordinate starts at 0, and x does not vary across the line segment, its value at the current level of *Dept*. Typically, create Cleveland dot plots in a horizontal direction. Obtain this orientation with `ggplot2` from the addition of the `coord_flip()` function.

3.1.3   Bubble Plot of Counts

A 1-dimensional scatterplot of the data values of a categorical variable is not appropriate for variables with a small number of unique values. The problem is *over-plotting*, in which a point is plotted many times at the same coordinate because the coordinate values repeat many times.

over-plot: A point is plotted multiple times in the same location.

One adjustment to over-plotting is the *bubble plot*. A bubble plot increases the size of the plotted point, the "bubble", according to the number of data values that plot to that respective coordinate. Figure 3.4 illustrates a 1-dimensional bubble plot for a categorical variable with five unique values, *Dept*.

bubble plot: Expand the size of a plotted point according to the value of another variable.

Because over-plotting renders the traditional scatterplot meaningless, the `lessR` scatterplot function, `Plot()`, by default, adjusts for over-plotting typical for a categorical variable by increasing the size of the corresponding plotted points. `Plot()` displays the corresponding count inside of each bubble large enough to accommodate the display. The `size.cut` parameter controls the number of values to display as text. To turn off the text display, set to `FALSE` or 0. Specify an integer larger than zero to display the indicated number of values. For example, the value of 2 plots the maximum and minimum values.

For `ggplot2`, `geom_count()` provides the bubbles based on counts, in conjunction with the scaling function `scale_size_area()`. The default size of the bubbles is

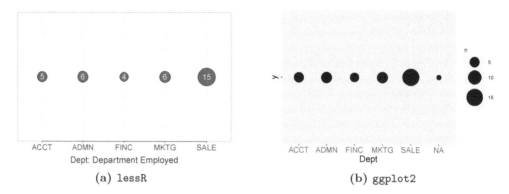

(a) `lessR`   (b) `ggplot2`

Figure 3.4: Bubble plots of counts for *Dept*.

somewhat small, so in this example, increase their maximum size with the `max_size` parameter.

> **R Input** *Bubble plot of Dept*
> *data*: d <- Read("http://lessRstats.com/data/employee.csv")
> ───
> *lessR*: Plot(Dept)
> *ggplot2*: ggplot(d, aes(Dept, "")) +
>   geom_count() + scale_size_area(max_size=14)

Scale the size of the bubble (circle) according to the corresponding value, such as the count, either via the radius of the circle or the circle's area. Human perception appears to assess area more accurately. The above `ggplot2` code explicitly specifies this scaling. With `lessR` this scaling based on a function of the area is implicit, with full control of the scaling available. The `radius` parameter, with a default of 0.25 inches, is equivalent to the `ggplot2 max_size` parameter. The `lessR` parameter `power` sets the relative size of the scaling of the bubbles to each other. The default value of 0.5 scales the bubbles so that the area of each bubble is the value of the corresponding sizing variable. The value of 1 for `power` scales such that the radius of the bubble equals the value of the sizing variable, increasing the discrepancy of size between the variables.

3.1.4 Display Proportions

Display either counts, also referred to as frequencies, or proportions, also referred to as relative frequencies, on the numeric axis.

Bar chart. The bar charts in Figure 3.2 display the proportions on the axis, as well as the rounded, corresponding percentages directly on the bars.

The bar charts in Figure 3.2 display without indication of missing data. The `lessR` function `BarChart()` automatically removes missing data in the display of the bar chart, and then reports the number of missing data values to accompany the frequency distribution displayed at the R console. By default, `ggplot2` includes the missing data as one of the categories on the categorical data axis. To optionally

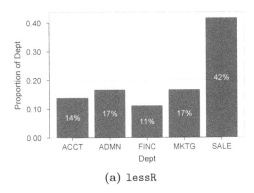
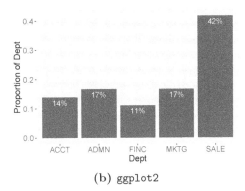

(a) lessR   (b) ggplot2

Figure 3.5: Bar chart of proportions from tabulated values with corresponding percentages displayed directly on the bars.

remove the missing data from the visualization, first process the data in the d data frame, such as with the base R function subset(), to retain only non-missing values of *Dept*. Here the base R function is.na() evaluates a data value for missing, used with the logical operator !, the R symbol for *not*.

BarChart statistical output, Listing 2.1, p. 30

R Input *Bar chart of proportions from tabulated values, no missing values*
data: d <- Read("http://lessRstats.com/data/employee.csv")

---

lessR: BarChart(Dept, stat="proportion")
ggplot2: d %>% subset(!is.na(Dept)) %>%
      ggplot(aes(x=Dept, y=(stat(count))/sum(stat(count)))) +
        geom_bar() + ylab("Proportion of Dept") +
        geom_text(aes(label=percent(stat(count)/sum(stat(count)),
              accuracy=1), y=(stat(count))/sum(stat(count))),
              stat="count", vjust=1.5, color="white")

To display proportions on the numerical axis, with lessR set the parameter stat to "proportion", which results in Figure 3.5a. To turn off the display of the percentage values on the bars, set values to "off". Or, display counts to complement the proportions on the numerical axis with values to "input".

For ggplot2, to display the proportions on the numerical axis, explicitly compute the values of y, the proportions. The geom_bar() function tabulates the counts as the values of y, so within the aes() function access the counts with the ggplot2 expression stat(count). Convert to proportions by dividing by their sum using the sum() function. Then the geom_bar() function plots the bars with the proportions on the corresponding axis.

Display the corresponding percentages on the bars with the geom_text() function. Invoke the percent() function from the scales package, part of the tidyverse. Here set the percent() parameter accuracy to 1 to round to the nearest whole number, as opposed to the default value of 0.1, which displays one decimal digit. Again, specify the computation of the proportions for their display, formatted as percentages, on the bars, as well as the y-coordinates of the displayed percentages.

Move the displayed percentages down into the bars themselves by setting the parameter `vjust` to 1.5. Positive values of `vjust` move the corresponding vertical coordinates down.

Cleveland dot plot. Figure 3.6 shows the `lessR` and `ggplot2` Cleveland dot plots with proportions displayed instead of counts as in Figure 3.3.

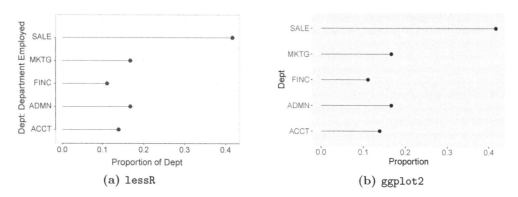

(a) `lessR`　　　　　　　(b) `ggplot2`

Figure 3.6: Cleveland dot chart of proportions from tabulated values.

Cleveland dot plot of counts, Section 3.1.2, p. 49

To display proportions with `lessR`, set `stat` to `"proportion"`. For `ggplot2`, follow the same basic construction for the Cleveland dot plot as with counts, except add the additional step of converting the computed variable *Freq* in the coordinate data table to relative proportions, here named *Prop*. Accomplish the transformation with the tidyverse function `mutate()` from the `dplyr` package.

R Input *Cleveland dot chart of proportions from tabulated values, no missing values*
data: d <- Read("http://lessRstats.com/data/employee.csv")

```
lessR: Plot(Dept, stat="proportion")
ggplot2: d %>% subset(!is.na(Dept)) %>% group_by(Dept) %>%
         summarise(Freq=n()) %>% mutate(Prop=Freq/sum(Freq)) %>%
      ggplot() +
        geom_point(aes(Dept, Prop), size=2.5) + coord_flip() +
        geom_segment(aes(x=Dept, y=0, xend=Dept, yend=Prop), size=.25)
```

The `lessR` `Plot()` also provides for displaying percentages. To do so, set `stat` to `"%"`.

3.2 Multiple Plots on a Single Panel

In many analyses, the interest is not only the bar chart or related visualization for a single variable but also to compare responses across multiple variables. A standard application is Likert scale data, common to social science research based on self-report attitude scales in which the respondent indicates the extent of agreement with one or more items on a survey or questionnaire. The most efficient display of multiple visualizations presents the comparative visualization on a single panel, illustrated in this section with the standard `lessR` functions `BarChart()` and `Plot()`.

An example of a Likert scale provides the following six unique data values based on response categories:

`Strongly Disagree, Disagree, Slightly Disagree, Slightly Agree, Agree, Strongly Agree`
Score the corresponding response categories with integer values, such as from 0 for Strongly Disagree to 5 for Strongly Agree. Assuming interval level responses, the values can be analyzed numerically, such as computing the mean and standard deviation. Because of the restricted set of responses, however, the analysis of the responses as categorical labels may be considered more appropriate.

interval data,
Section 1.2.3, p. 10

The Mach IV (Christie & Geis, 1970), which measures the tendency to agree with Machiavellian attitudes, the manipulation of others for selfish, personal gain. The R console displays the content of these items if included as variable labels. Figure 3.7 summarizes the responses of 351 respondents to these items from the Deceit sub-scale (Hunter, Gerbing, & Boster, 1982).

variable labels,
Chapter 1.2.5, p. 18

```
                       ──────── four machiavellian items ────────
1  m06: Honesty is the best policy in all cases
2  m07: There is no excuse for lying to someone else
3  m09: All in all, it is better to be humble and honest than to be important and
4       dishonest
5  m10: When you ask someone to do something for you, it is best to give the real
6       reasons for wanting it rather than giving reasons which carry more weight
```

multiple bar charts, one panel, Figure 2.3, p. 33

Chapter 2, the Quick Start chapter, introduced the visualization of stacked bar charts, each bar chart occupying a single column, illustrated with the full 20-item Mach IV scale. For comparison, Figure 3.7 visualizes only the four items of the Deceit sub-scale. The comparison is to the visualization called here the Bubble Plot Frequency Matrix (BPFM). This visualization extends the bubble plot to many variables simultaneously, which typically, but not necessarily, share the same response scale.

Mach IV Likert items, Section 1.2.4, p. 17

Figure 3.7 illustrates the stacked bar chart and BPFM of these four Likert-scaled items.

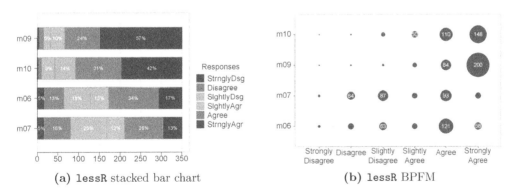

(a) `lessR` stacked bar chart **(b)** `lessR` BPFM

Figure 3.7: Single panel plots of responses to multiple variables.

To generate the BPFM, pass a vector of categorical variables to `Plot()` instead of a single variable. The result is a bubble plot for each variable in the vector, with all the bubble plots stacked on top of each other. In R the most general way to specify

c() function: Combine a set of objects into a single object.

a vector is with the `c()` function. Here define the vector as a list of variable names, each name separated by a comma.

`Plot()` recognizes if a variable is numeric or character, and the subsequent analysis and visualization proceeds accordingly. The responses to the Likert items are numeric, recorded as integers from 0 to 5. To analyze the Likert responses as character values, one possibility first converts each categorical variable to an R factor, such as with lessR `factors()`. In general, before beginning an analysis, convert categorical variables, either with integer or non-numerical values, to R factors.

R factors,
Section 1.2.6, p. 19

create factor variable
with `factors()`,
Section 1.2.6, p. 21

To maximize the readability of the output, provide meaningful value labels, the labeled response categories, when converting the numerical responses for each variable (item) to a factor. In this example, store the labels in the vector LikertCats, and then convert with the lessR function `factors()`, which converts all 20 Mach IV items to factors with a single function call.

R Input *Default Bubble Plot Frequency Matrix with Value Labels*

```
data: d <- Read("http://lessRstats.com/data/Mach4.csv")
   LikertCats <- c("Strongly Disagree", "Disagree", "Slightly Disagree",
               "Slightly Agree", "Agree", "Strongly Agree")
   d <- factors(m01:m20, levels=0:5, labels=LikertCats)
```

lessR: BarChart(c(m06,m07,m09,m10))
lessR: Plot(c(m06,m07,m09,m10))

An alternative to creating a factor for each item invokes the lessR parameter `n.cat`, an abbreviation for the "number of categories" that specifies the largest number of unique, equally spaced integer values of a variable to define categorical instead of numerical responses. The default is 0, that is, analyze values with equally-spaced integer values as numeric. In this example set `n.cat` to 6 to indicate up to six unique, equally spaced values to analyze as categorical. If specified, can also present the content labels for each category with the lessR parameter `value.labels`.

The series of stacked bar charts or BPFM's provide a more compact display of the responses to a set of response scales than does the corresponding display of individual bar charts or bubble charts. Further, the responses across the different items compare more easily with the BPFM.

3.3 Provide the Numerical Values

One option inputs the *y*-values directly. In this first example, directly plot a subset of the original data table of measurements, where each row represents the observations on a single unit, such as a person, company, or geographical region. That is, some of the original data values become the coordinates of the bar chart or related visualization.

3.3.1 Bar Chart of Individual Data Values

From the employee data table referenced throughout this book, Figure 3.8 presents a bar chart of test scores of employees before instruction, the variable *Pre*. Limit the visualization to ten employees to conserve space. Plot the names and the test scores as the existing data values in the data table.

employee data table, Section 1.2.2, p. 7

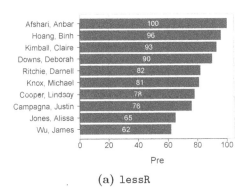
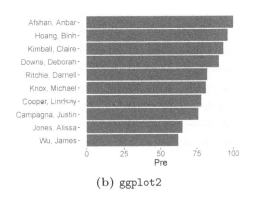

(a) `lessR`   (b) `ggplot2`

Figure 3.8: Bar charts for individual data values.

After reading the data into the *d* data frame, retain only the first ten rows of the data, here with the base R `Extract` function. Indicated by the square brackets, the extraction retains all variables, as no information appears after the comma in the `Extract` expression.

Extract function, Section 1.2.4, p. 13

This example illustrates the concept of *row names* in standard base R data frames. The row name labels each row with a unique identifier, sequential integers by default. Each employee's name serves as that unique identifier. Or, the names can be read as a separate variable, though their primary purpose is as an identifier, not a variable for analysis. Create row names with the base R parameter `row.names` set to the corresponding column number as part of the call to a read function such as `Read()`, as in the following example. Or, create the row names following the read with the base R function `row.names()`.

row names of a data frame: Unique name of each row of a data frame, not a standard variable.

The tidyverse concept of a data frame, the *tibble*, does not include row names beyond numbered rows. Information such as employee names stored in the row name column of a standard R data frame necessarily stores in a tibble as another variable. Further, to transform a standard data frame with a tidyverse function, such as transforming a variable with `mutate()`, the processing strips the row names from the original data frame.

If a data frame contains row names, can they be analyzed as a variable? For `lessR`, access a data frame's row names in the visualization functions as the variable `row.names`, treated as any other variable. For the tibble, to preserve the row name information, before processing with a tidyverse function invoke either the base R function `row.names()` to create a new variable from the row names, or use the `as_tibble` function with the parameter `rownames_to_column`.

R Input *Bar charts for individual data values*

```
data: d <- Read("http://lessRstats.com/data/employee.csv", row.names=1))
      d <- d[1:10,]
```

```
lessR: BarChart(row_names, Pre, horiz=TRUE, sort="+")
ggplot2: as_tibble(rownames_to_column(d, var="Name")) %>%
             arrange(Pre) %>%
             mutate(Name = factor(Name, levels=Name)) -> tbl
          ggplot(tbl, aes(Name, Pre)) + geom_col() + coord_flip() +
             theme(axis.title.y=element_blank())
```

To create the visualizations, for `lessR` specify `row_names` as the variable for parameter x and the numerical variable to plot as the parameter y, then flip the coordinates by setting `horiz` to TRUE. Sort the output from most to least by setting `sort` to `"+"`.

For `ggplot2`, accomplish the data manipulations before the call to `ggplot()`. Sort the data with the `arrange()` function. Then convert the variable *Name* to a factor to preserve the ordering. Assign the results of the transformations to the tibble *tbl* with the assignment arrow written as `->`, to indicate the direction of the assignment. Pass the transformed data to `ggplot()` and remove the self-evident axis label for `Name` from the visualization.

3.3.2   Vertical Long Value Labels

As illustrated in the preceding examples, for the horizontal orientation, the `lessR` and `ggplot2` bar charts automatically adjust the left margin to accommodate large value labels, in this case, employee names. With a vertical orientation, however, there may not be enough room to accommodate the width of the category labels. With a vertical bar chart, both `lessR` and `ggplot2` default to a horizontal display of the category labels, which for long value names results in a jumbled mess of overlapping values.

One solution to the problem of long labels rotates the value labels 90 degrees to display vertically. With that modification, the bottom margin adjusts by default to accommodate the value labels, as shown in Figure 3.9.

Rotate the labels for `lessR` with the `rotate_x` parameter, here set to 90. For `ggplot2`, invoke the `element` function with the `angle` parameter, here set to 90, embedded within the `theme` function. With such a large number of bars to display, by default `lessR` does *not* display the value for each category within each corresponding bar, that is, `values` set to `"off"`.

%>% pipe operator, Section 2.5.1, p. 41

Unlike the previous example, here do not read employee names as row names but rather as an ordinary variable, *Name*, according to the first row of variable names in the referenced `csv` data table. Also, the pipe operator, `%>%`, transforms the read data frame, *d*, and then feeds the prepared data directly into `ggplot()`.

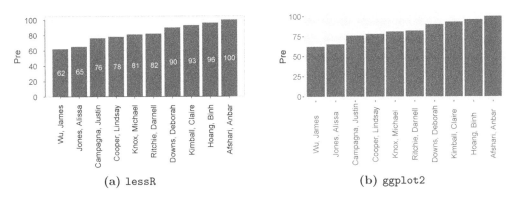

(a) `lessR` (b) `ggplot2`

Figure 3.9: Bar charts for individual data values.

R Input *Bar chart with x-axis values rotated 90 degrees*

data: d <- Read("http://lessRstats.com/data/employee.csv")
        d <- d[1:10,]

```
lessR: BarChart(Name, Pre, sort="+", rotate_x=90)
ggplot2: d %>% arrange(Pre) %>%
         mutate(Name = factor(Name, levels=Name)) %>%
       ggplot(tbl, aes(Name, Pre)) + geom_col() +
         theme(axis.text.x=element_text(angle=90, vjust=0.5)) +
         theme(axis.title.y=element_blank())
```

By default, `BarChart()` provides some help for long value labels with a vertical bar chart. If the value labels contain spaces, which indicate separate words, and if there is no rotation specified, then the function displays each new word on a new line.

3.3.3 Cleveland Dot Plot of Individual Data Values

Figure 3.10 shows the Cleveland dot plot that replaces the bar chart in Figure 3.8. As with the previous example of a Cleveland dot plot of tabulated data that

Cleveland dot plot of tabulated data, Figure 3.3, p. 48

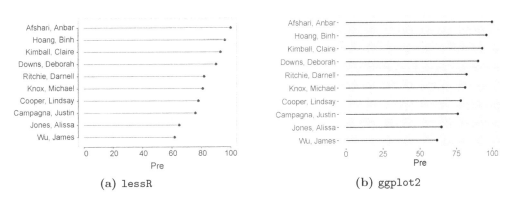

(a) `lessR` (b) `ggplot2`

Figure 3.10: Cleveland dot plots for individual data values.

replaced a bar chart, a Cleveland dot plot of individual values can also replace the corresponding bar chart of individual values.

For both `lessR` and `ggplot2`, specify the start point of the numerical axis if it is to begin at a value other than the minimum data value of 62 for the Pre test score. With `lessR`, set the parameter `origin.x` to 0. For `ggplot2`, set the starting value as the y parameter in the call to `geom_segment()`.

R Input *Cleveland dot plots for individual data values*

```
data: d <- Read("http://lessRstats.com/data/employee.csv"))
      d <- d[1:10,]
```

```
lessR: Plot(Pre, Name, origin.x=0)
ggplot2: d %>% arrange(Pre) %>%
             mutate(Name = factor(Name, levels=Name)) %>%
         ggplot(aes(Name, Pre)) + geom_point() + coord_flip() +
             geom_segment(aes(x=Name, y=0, xend=Name, yend=Pre)) +
             theme(axis.title.y=element_blank())
```

The remaining code to generate the Cleveland dot plots in Figure 3.10 follows the same pattern as dot plots that tabulate the counts for each of the levels of the categorical variable.

3.3.4   Visualize Means across Categories

Given an original data table of measurements with categorical variable x and numerical variable y, obtain the coordinate data table from which to construct the visualization by summarizing the values of y for each level of x. The example here is the employee data table, with categorical variable *Dept* and numerical variable *Salary*. Given the original data table of measurements, visually display the sample mean of *Salary* for each level of *Dept*.

employee data table,
Section 1.2.2, p. 7

Dept	Avg
ACCT	$51,792.78
ADMN	$71,277.12
FINC	$59,010.68
MKTG	$60,257.13
SALE	$68,830.06
<NA>	$43,772.58

Figure 3.11: Mean salary by department.

coordinate data table,
Figure 3.1, p. 46

The coordinate data table in Figure 3.11 summarizes the original data table of measurements with mean values of *Salary*, named *Avg*, for each level of *Dept*. The `<NA>` from Figure 3.11 indicates that the department of employment for at least one employee is missing from the original employee data table.

Bar chart

Figure 3.12 presents the default `lessR` bar chart and the `ggplot2` bar chart constructed from the summarized data, both with added displayed values on the bars from the data in Figure 3.11. Again, without first setting the color theme to

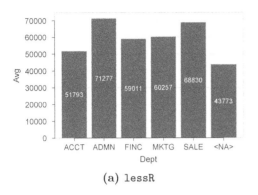 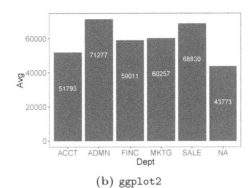

(a) `lessR`   (b) `ggplot2`

Figure 3.12: Bar charts direct from the data values.

grayscale with `style("gray")`, the bars on the `lessR` bar chart display a variety of hues with the same chroma and luminance according to the `lessR` `"colors"` palette based on the HCL color space.

`"colors"` palette, Section 10.2.1, p. 215

The `lessR` function `BarChart()` provides the option of internally computing the coordinate data table from the original measurements with the `stat` parameter for one of a variety of summary statistics. Valid values consist of `"mean"`, `"sum"`, `"sd"` for standard deviation, `"dev"` for mean deviation, `"min"`, `"median"`, and `"max"`. In the call to `BarChart()`, specify the x and y parameters, respectively, by first listing the two variable names, here *Dept* and *Salary*, then the `stat` parameter with the specified summary statistic.

hue, chroma, luminance, Section 10.1.1, p. 212

For the corresponding function that specifies the variables to plot, `BarChart()` for `lessR` or `aes()` for `ggplot2`, the first position in the parameter list defaults to the value of the categorical variable x, and the second position for the numerical variable y. Given the coordinate data table as the input data, to inform `ggplot2` *not* to tabulate the values of x to obtain the values of y, but instead to use the provided values of y, set the `stat` parameter to `"identity"` in the call to `geom_bar()`. Or, invoke the `geom_col()` function that automatically sets `stat` to `"identity"`. For `lessR`, if the function call includes a value of y, then plot its values. For both `lessR` and `ggplot2`, because y is the second parameter in the list of parameters, in the function call the value for the parameter y can be unnamed if listed as the second parameter.

geom_col() function, `ggplot2`: Plot data values as bars directly entered without transformation.

For `ggplot2`, to calculate the coordinate data table in Figure 3.11 from the original employee data table, here use the tidyverse `dplyr` functions, `group_by()` to define the groups and `summarize()` to calculate summary statistics, here the `mean` of *Salary*. To retain the original or raw data in the d data frame, create a separate data frame, *avgd*, to store the summarized data. To analyze data from a data frame other than the `lessR` default of d, apply the `data` parameter.

%>% pipe operator,
Section 2.5.1, p. 41

This example applies the **magrittr** package pipe operator, **%>%**, to transfer the output from one function forward to the value of the first parameter value for the next function in the sequence, here the input to that function. Either load the **magrittr** package separately with **library()**, or load as part of the tidyverse, with all of the tidyverse packages and associated functions.

> **R Input** *Bar chart direct from provided values of mean* Salary *with labels*
> *data*: d <- Read("http://lessRstats.com/data/employee.csv")
>
> ---
>
> *lessR*: BarChart(Dept, Salary, stat="mean")
> *ggplot2*: d %>% group_by(Dept) %>% summarize(Avg=mean(Salary)) -> avgd
>   ggplot(avgd, aes(Dept, Avg)) + geom_bar(stat="identity") +
>   geom_text(aes(label=round(Avg,0)), vjust=8, size=4.2,
>   color="white")

values, `lessR`:
Parameter to specify
the display of values
on the bars of a bar
chart.

By default, the **lessR** bar chart includes the plotted data values, the values of y, on each bar. To customize the display of the values set the **values** parameter as **"off"**, **"%"**, **"proportion"**, or **"input"**. For y as tabulated counts, **values** defaults to **"%"**, which lists the percentage of responses for each category, the relative frequency. For y provided directly, as in this example, **value** defaults to **"input"**, the values of y from which the analysis proceeds. The value of **"proportion"** displays the proportions instead of percentages.

Related parameters are **values_color** for the color of the displayed values, **values_size** for their size, **values_digits** for the number of decimal digits, **values_pos** for position inside or outside of the bars (**"in"**, the default, or **"out"**), and **values_cut**, which specifies the minimum value for which to provide a display.

BarChart() also allows the separate computation of the coordinate data table in Figure 3.11, and then enter the coordinate data table directly, as is done with **ggplot2**.

> `BarChart() function call to process coordinates directly`
>
> ---
>
> *lessR*: BarChart(Dept, Avg, data=avgd)

Instead of analyzing *Salary* from the default data frame *d*, analyze the average mean salary, *Avg*, from data frame *avgd*.

For **ggplot2**, include the means for each group displayed on the bars with the **geom_text()** function. The **vjust** parameter shifts the vertical position of each displayed mean, with positive value shifting downward. The **size** and **color** parameters provide the expected information. The values of these three parameters for the function calls that generated the **ggplot2** bar chart in Figure 3.12 were obtained with trial and error over successive iterations of generating the bar chart.

Cleveland dot plot

Figure 3.13 displays the Cleveland dot plot version of the bar chart of the mean *Salary* across departments. In this example, the categories remain in alphabetical order instead of sorted by average *Salary*. The starting point on the numerical axis is \$50,000.

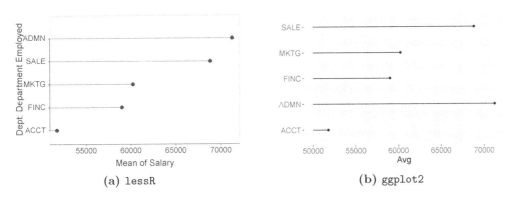

(a) `lessR`   (b) `ggplot2`

Figure 3.13: Cleveland dot plots for mean of *Salary* across levels of *Dept*.

stat `lessR` parameter, Section 3.3.4, p. 59

For `lessR`, the syntax remains the same as for the bar chart, except invoke `Plot()` instead of `BarChart()`. The same options for the `stat` parameter are available, including `"mean"`, invoked in this example. The `ggplot2` function calls also follow the same pattern as previous examples. The sequence of data transformations culminates in the creation of the data frame *avg*, indicated by the assignment arrow `->`.

R Input *Dot plot of mean* Salary *by* Dept

data: `d <- Read("http://lessRstats.com/data/employee.csv")`

```
lessR: Plot(Dept, Salary, stat="mean")
ggplot2: d %>% subset(!is.na(Dept)) %>%
         group_by(Dept) %>% summarize(Avg=mean(Salary)) -> avgd
       ggplot(avgd, aes(Dept, Avg)) + geom_point() + coord_flip()
         geom_segment(aes(x=Dept, y=50000, xend=Dept, yend=Avg))
```

The next example illustrates some options in the analysis of mean deviations across the categories.

3.4   Communicate with Bar Fill Color

With the previous examples of bar charts in this chapter, the bars communicated only length and so could be replaced by Cleveland dot plots. In this section fill the bars with various colors to communicate information regarding the underlying data. The first example continues from the last section on visualizing the mean of *Salary* across the five departments of a company. Here instead of the mean, focus on the mean deviation for each department.

3.4.1 Bar Fill Color Bifurcated by Value of Mean Deviations

Some analyses involve the comparison of values against a baseline value, such as zero. To emphasize the contrast between values on either side of the baseline, construct the bar graph with the bars of negative values filled with one color and positive values filled with another color to emphasize the contrast. The colors can be the same for each set, or the colors can blend into each other. First, consider two distinct colors.

(a) `lessR`   (b) `ggplot2`

Figure 3.14: Bar charts with different colors for negative and positive values.

<u>Two distinct colors</u>. The example follows a further transformation of the average salary data from the previous section. To focus on the differences between mean salaries across the departments, plot and compare their deviations from the overall (unweighted) mean. In Figure 3.14 highlight the negative vs. positive values with a light shade of gray compared to a darker shade, though in general, choose more dramatic contrasts such as red vs. blue or black.

To visualize the direct difference of each department's salary compared to the mean, first compute the mean deviations, here stored in the *avgd* data table shown in Figure 3.15. Remove the row of missing data indicated in Figure 3.11 with the base R `na.omit()` function, which removes all rows of data from a data frame that contain any missing data. Calculate and then store the mean deviations in the newly created variable *Dev*. To

Dept	Avg	Dev
ACCT	$51,792.78	-$10,440.78
ADMN	$71,277.12	$9,043.57
FINC	$59,010.68	-$3,222.88
MKTG	$60,257.13	-$1,976.42
SALE	$68,830.06	$6,596.51

Figure 3.15: Data table *avgd*, mean deviations.

specify a variable within a data frame, include the name of the data frame, followed by a $, and then the variable name.

different strategy for ggplot2 sorting, Section 6.2.1, p. 141

Two different strategies to obtain these bar charts in Figure 3.14 follow from `lessR` and `ggplot2`. The `BarChart()` parameter `fill_split` indicates to create the bar graph with two different fill colors, split according to the provided value of *y*. For `ggplot2`, create a binary variable that indicates the status of the corresponding value of *y*, negative or zero, or positive, and then map these values into the `fill` parameter.

Also, `BarChart()` provides `stat` with the value of `"dev"` to internally compute the coordinate data table of mean deviations. With `ggplot2`, manually calculate the coordinate data table.

R Input *Bar chart with different colors for negative vs. positive values*
data: d <- Read("http://lessRstats.com/data/employee.csv")

---

lessR: BarChart(Dept, Salary, stat="dev", sort="+", fill_split=0)
ggplot2: avgd <- d %>% subset(!is.na(Dept)) %>%
        group_by(Dept) %>% summarize(Avg = mean(Salary)) %>%
        mutate(Dev = Avg - mean(Avg)) %>%
        mutate(neg.pos = ifelse (Dev <= 0, "-", "+"))
      ggplot(avgd, aes(reorder(Dept, Dev), Dev, fill=neg.pos)) +
        geom_col() + scale_fill_grey(start=.3, end=.7)

The `BarChart()` `sort` parameter sorts the presentation of the bars according to the value of *y*, *Dev* in this example. The value of `"+"` sorts in ascending order. For `lessR`, the default fill colors are light and dark shades from the first color from the qualitative palette, `"colors"`, blue. This example relies upon the default gray colors from the gray color theme set by `style("gray")`. To override the default for any color theme specify a `fill` vector of two colors, such as `c("darkred", "gray30")` to plot a shade of red for negative values and a dark gray for positive values.

qualitative palette, Section 10.2.1, p. 216

For `ggplot2`, here name the created variable *neg.pos*. Assign the values of neg.pos with the base R function `ifelse`, such as `"-"` if negative or zero, and `"+"` if positive. To sort the bar presentation of *Dept* according to the order of *Dev*, use the base R function `reorder()` within `aes()` that specifies the variable names. Map the values of *neg.pos* to the fill color of the bars by setting `fill` to *neg.pos*. By default, the two displayed fill colors are the first two colors from `ggplot2` qualitative scale, red and blue. To display in grayscale, the specialized `scale_fill_grey()` function would typically be used, but with the more general `scale_fill_manual()` the visualization can extend to any two colors.

map a variable to visual aesthetic, Section 10.2.1, p. 217

<u>Two colors that blend into each other</u>. Another option specifies two bar colors as a divergent color palette so that one color blends into the other as the bars progress from the smallest value of *y* to the largest value. Typically define the divergent color palette over two distinct colors, such as red and black. For purposes of illustration in Figure 3.16, to maintain grayscale specify the same value of gray for both colors in this example.

divergent color palette, Section 10.2.3, p. 221

The advantage of `ggplot2` for this application is that its colors are mapped to the `fill` parameter according to the corresponding value of *y*. `BarChart()` presents a divergent color palette, but the progression only relates to the bar color according to the ordinal position of the bar in the progression, not its numerical value.

For `lessR`, define a divergent color palette with an implicit call to the `lessR` function `getColors()`, which relies upon `diverge_hcl()` from the `colorspace` package (Ihaka, Murrell, Hornik, Fisher, & Zeileis, 2016). Specify a vector of two plural color names from the 12 hues of the HCL color wheel in Figure 10.3. Or, as

HCL color wheel, Figure 10.3, p. 212

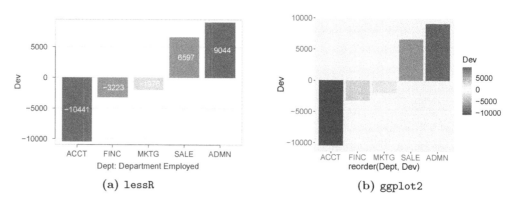

Figure 3.16: Bar charts with a divergent palette of colors for negative and positive values.

in this example, confine the palette to grayscale with `"grays"`. For `ggplot2`, define the divergent gradient with the function `scale_fill_gradient2()`, specifying a `low` value and a `high` value. This function allows for even more control than `diverge_hcl()` because it provides a `midpoint` parameter so that the neutral midpoint can be explicitly specified.

> **R Input** *Bar chart with diverging gradient for negative vs. positive values*
> *data*: `avgd` data frame continued from previous example
>
> ---
>
> *lessR*: `BarChart(Dept, Salary, stat="dev", sort="+",`
> `        fill=c("grays", "grays"), color="off", values="input")`
> *ggplot2*: `ggplot(avgd, aes(reorder(Dept, Dev), Dev, fill=Dev))`
> `            geom_bar(stat="identity") +`
> `            scale_fill_gradient2(low="gray25", high="gray25", midpoint=0)`

In general, specify two distinct colors for the divergent palette in place of grays. For `lessR`, as indicated, choose two of the 12 hues defined across the HCL color wheel from Figure 10.3 in place of the `"grays"`, such as `"greens"` and `"violets"`. The result is a color palette generated from HCL colors. Or, call `getColors()` directly as the argument to `"fill"`, which also permits the specification of chroma `c` and luminance `l` at custom values. For even more control, refer directly to `diverge_hcl()` from the `colorspace` package, which `lessR` relies upon to generate a divergent palette.

getColors()
function,
Section 10.2.1, p. 215

For `ggplot2`, substitute any color reference for the `"gray25"` instances in the previous example including HCL colors. One example, which transforms from a vibrant red color to a vibrant blue color, specifies `low` as `hcl(0,90,40)` and `high` as `hcl(240,90,40)`.

hcl() function,
Section 10.1.1, p. 212

3.4.2   Bar Chart of an Ordinal Variable

Ordinal variable,
Section 1.2.3, p. 10

Ordered factor,
Section 1.2.6, p. 21

The ordered values of an ordinal variable do not represent numerical values. How to portray the fill color of the bars according to the ordinal position of the corresponding level? One solution maps the discrete values into a color range according to the position of each value along the ordered range. To illustrate, consider the variable from the employee data *JobSat* with character string values of "low", "med" and "high". To define the variable as ordinal within R, transform *JobSat* from character strings as initially read into R into an ordered factor.

The ggplot2 default palette for mapped values of an ordered factor to the fill and color parameters is viridis, so the bars of a bar chart of an ordered factor display with viridis colors without any additional specifications. For lessR, specify viridis as the fill color.

virdis palette,
Section 10.2.2, p. 220

> **R Input** *Bar chart of Counts with viridis colors for an ordered factor*
> *data*: d <- Read("http://lessRstats.com/data/employee.csv")
>   d <- factors(JobSat, levels=c("low", "med", "high"), ordered=TRUE)
>
> *lessR*: BarChart(JobSat, fill="viridis")
> *ggplot2*: ggplot(d) + geom_bar(aes(JobSat, fill=JobSat))

The viridis colors do not show here, but grayscale does. Figure 3.17 illustrates the bar chart of the ordered factor *JobSat* with a sequential grayscale.

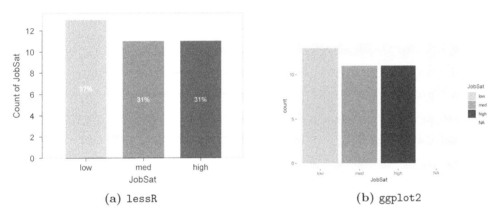

(a) lessR   (b) ggplot2

Figure 3.17: Bar charts of an ordered factor filled with a sequential grayscale.

For lessR, to generate Figure 3.17, the fill parameter could be set to "grays", but the entire theme, as is true with all the lessR illustrations in this book, has been set to "gray" with the style function. lessR automatically chooses the hue for a sequential color palette by default according to the current theme. For example, if the theme had been "dodgerblue", then the pre-defined palette would be "blues".

ggplot2 defaults to colors on a blue scale. Obtain the grayscale with the scale_fill_grey() function, here with the bar for the missing data set to "white" according to the parameter setting for na.value. Specify start and end values for the shades of gray.

> **R Input** *Bar chart of Counts with a sequential gray scale for an ordered factor*
> *data*: from previous example
>
> *lessR*: BarChart(JobSat)
> *ggplot2*: ggplot(d) + geom_bar(aes(JobSat, fill=JobSat)) +
>   scale_fill_grey(start=.7, end=.2, na.value="white")

HCL color wheel,
Figure 10.3, p. 212

To leave the defaults behind and leave grayscale behind, specify another color palette, here the `lessR` palette `"greens"` from the HCL color wheel.

> **R Input** *Bar chart of Counts with a green sequential scale for an ordered factor*
> *data*: from previous example
>
> ---
> *lessR*: `BarChart(JobSat, fill="greens")`
> *ggplot2*: `ggplot(d) + geom_bar(aes(JobSat, fill=JobSat)) +`
> `        scale_fill_manual(values=getColors("greens", n=3),`
> `            na.value="gray50")`

In this example, use the `lessR` function `getColors()` to generate three green colors from light to dark for both the `lessR` and **ggplot2** bar charts with the green colors mapped into the bar colors.

Map Continuous Values to Fill Color

map data to `fill`,
Section 10.2.1, p. 217*

*map data to the size
aesthetic,* Figure 5.10,
p. 115

Mapping values of a variable to parameter `fill` for the discrete values of a categorical variable relates colors to those values. Here map the values of a continuous variable to the fill color of the bars. Instead of a palette defined by pre-selected values for different levels of the categorical variable, base the displayed colors on the value of a continuous variable, scaled according to the corresponding data values. Data may be mapped to other aesthetics as well, but here illustrate fill color.

For a bar chart, map the values for the numerical variable plotted on the numerical axis, y, to the `fill` aesthetic. Specify the values of y as part of the input data, or as shown in Figure 2.1, tabulate as the counts of the occurrence of each level of the categorical variable. Figure 3.18 illustrates this mapping of tabulated counts to the bar fill that plots the bars for the larger counts with colors.

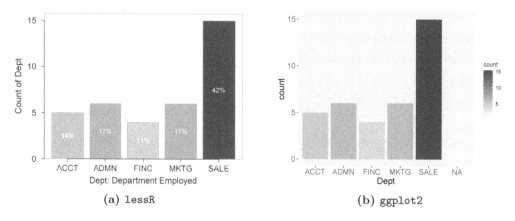

Figure 3.18: Map the value of the count to `fill`.

For the plot of the tabulated counts, y is not specified, so provide the means for the values of y. For `lessR`, to map into `fill`, specify the name (`count`), deliberately chosen as an invalid R name so as not potentially conflict with the name of an existing variable. For **ggplot2**, invoke the `stat()` function with the value of `count`, embedded in `aes()`. `ggplot2` by default also lists the number of missing values as part of the plot (as opposed to listing at the console as does `lessR`).

> **R Input** *Default bar chart of Counts*
>
> *data*: `d <- Read("http://lessRstats.com/data/employee.csv")`
> ---
> *lessR*: `BarChart(Dept, fill=(count))`
> *ggplot2*: `ggplot(d, aes(Dept, fill=stat(count))) + geom_bar()`

As with all `lessR` examples in this book, achieve the `lessR` grayscale by setting the theme to `"gray"` with `style()`. Changing the theme for `ggplot2`, does not change the bar color, for which the default sequential scale is a range of blue colors. To obtain the grayscale for the bars turn to the function `scale_fill_gradient()`, which creates a gradient across a range of colors from the `low` color to the `high` color.

scale_fill_gradient() function, `ggplot2`: Create a gradient of colors from low to high.

> *ggplot2*: `...  scale_fill_gradient(low="gray95", high="gray25")`

To specify a general color progression, set the values of `low` and `high` accordingly. To change from the default blue sequential scale, for example, specify HCL colors with the base R function `hcl()` hue set to 0 for red, chroma to 85 and vary luminance from as light as 85 to as dark as 30. That is, set `low` to `hcl(0,85,85)` and set `high` to `hcl(0,85,30)` for a sequential red color scale that varies only in luminance. Or, vary the hues as well, mixing colors on the same gradient.

To obtain the viridis palatte for the bar fill, for `ggplot2`, transform the cateogrical variable x to an ordered factor. Then mapping y to bar fill as in the above example applies the viridis palette. For `lessR`, without a separate scaling function, specify the value of `(count.v)` for `fill` to obtain the viridis palette instead of the blue sequential palette.

3.4.3 Custom Color for Individual Bars

With the `lessR` and `ggplot2` `fill` parameter, specify any vector of colors for the individual bars. For example, highlight an individual bar. One application is a bar chart for Likert scale response data that includes a No Opinion option. Display the bars on the Disagree/Agree continuum with a sequential color palette such as from light colors to progressively darker colors to indicate higher levels of agreement for the corresponding attitude item. The No Opinion item, however, is not part of that continuum, so fill with a different color than from the chosen sequential palette.

The example in Figure 3.19 illustrates a bar chart with only two responses for item X on the Disagree/Agree continuum, Disagree and Agree. The bars for these responses are colored light to dark gray. The No Opinion option bar, colored white, follows the two other bars.

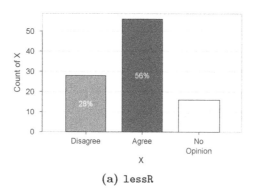
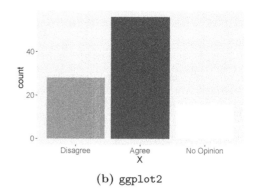

(a) `lessR`   (b) `ggplot2`

Figure 3.19: Custom color the No Opinion bar separately from the sequential progression of the gray scale bars.

Simulate the data for this example with randomly generated values. First, establish the values from which to sample the character vector *X.values*. The use of the base R repetition function, `rep()`, generates a vector with the specified number of repetitions of each value. Here the *X.values* vector consists of two values of Disagree, four values of Agree, and one No Opinion. Sampling from this vector will, on average, yield about twice as many Agree responses than Disagree, and four times as many No Opinion responses.

factor function,
Section 1.2.6, p. 19

Sample the values of X with the base R function `sample()`. The function generates a sample of 100 simulated responses of Disagree, Agree and No Opinion values. Then convert these values to an R factor with the base R `factor()` function to order the values as specified. The result is the vector X in the user workspace, or global environment, not part of any data frame. For `ggplot2`, specify `NULL` for the data frame to select the variable X from the user's workspace where X was created. `lessR` automatically searches the user's workspace for the variable.

> **R Input** *Custom color the No Opinion bar*
>
> *data*: X.values <- c(rep("Disagree",2), rep("Agree",4), "No Opinion")
>   X <- sample(X.values, size=100, replace=TRUE)
>   X <- factor(X, levels=c("Disagree", "Agree", "No Opinion"))
> ---
> *fill colors*: Clr <- c(getColors("grays", n=2), "white")
> *lessR*: BarChart(X, fill=Clr, color="black")
> *ggplot2*: ggplot(NULL, aes(X)) + geom_bar(fill=Clr)

The vector *Clr* contains the three fill colors. The `lessR` function `getColors()` generates the two grayscale colors that correspond to the first two bars, the Disagree and Agree bars. The complete *Clr* vector also includes the color `"white"`. Because the `lessR` bar charts have default light backgrounds, change the border of the bars from the default `"transparent"` to `"black"` to better display the white bar. For both `lessR` and `ggplot2`, specify the fill colors of the three bars with the *Clr* vector.

3.5 Create a Report from Saved Output

As illustrated in Chapter 2, `BarChart()` both creates the visualization and writes the statistical output as text to the R console[1]. `BarChart()` output can also be saved into an R list object, along with other statistical output. To do so, assign the output to a named R object, such as `b`.

BarChart()
statistical output,
Listing 2.1, p. 30

R Input *BarChart() saved output*

data: d <- Read("http://lessRstats.com/data/employee.csv")

lessR: b <- BarChart(Dept)

The saved `BarChart()` output is an R object of type `list`, which consists of different components. As with any R object, enter its name, *b*, at the console to view its contents. There are two sets of components for this saved list.

The first set consists of `out_title`, `out_lbl`, `out_counts`, `out_chi`, and `out_miss`. These components as a group define the text output of `BarChart()` at the console, but each exists separately. Respectively, these components define the title, any variable label, the frequency distribution, the chi-square test, and the number of missing values.

The second set of components represent output that can be input into further analysis, such as for input into other functions, or individually accessed for subsequent reports. These components consist of `n.dim`, `p_value`, `freq.df`, `freq`, `prop`, and `n.miss`. Respectively these components indicate the number of dimensions (variables), the *p*-value of the chi-square test, a data frame of the frequency distribution, a table of the frequencies, a table of the proportions, and the number of missing values.

The format of the following R Markdown document provides a template to embed the saved output of `BarChart()` into a report, which includes the presentation of results and their analyses. Find this R Markdown document online at `lessRstats.com/Rmd/04.Rmd`.

──────── Illustrative BarChart R Markdown document ────────

```
1   ## Analyze Number of Employees by Department
2   Load `lessR`, which includes the data.
3   ```r
4   suppressPackageStartupMessages(library(lessR))
5   d <- Read("Employee", quiet= TRUE)
6   ```
7
8   ### The Bar Chart
9   First run the `BarChart` function. Do not display statistics at the console.
10  ```r, fig.width=5.5, fig.height=3.5
11  b <- BarChart(Dept, quiet=TRUE)
12  ```
13
14  ### Statistics
15  The number of missing values is `r b$n.miss`. Obtain the frequency distribution.
16  ```{r, echo=FALSE}
```

---

[1]No output to the console if `quiet` is set to `TRUE`, either in the call to `BarChart()` or persistently in a call to `style`

```
17   print.out(b$out_counts)
18   ```
19
20   Obtain the chi-square test, for one variable with the null hypothesis of equal
21   population proportions.
22   ```{r, echo=FALSE}
23   print.out(b$out_chi)
24   ```
25
26   Reject the null hypothesis with a _p_-value of `r round(b$p_value,3)`. The number
27   of employees across departments differs in the population as well as the sample.
```

As with all R Markdown documents, embed lines of R code, a code block, between the line ```` ```{r} ```` and the closing line ```` ``` ````, entered manually. Or, more conveniently define a code block from the RStudio Insert drop-down menu in the toolbar above the source window. All other lines in this R Markdown document are commentary. Lines 1, 8, and 14 that begin with a pound sign, #, are headings. Two pound signs indicate a second-level heading and three pound signals a third level heading, which dictates the font size of the heading.

Refer to a specific component in the saved list object with the list name, followed by a $, and then the component name. This report contains two list components of formatted output in Lines 17 and 23, b$out_counts and b$out_chi, the frequency distribution and associated chi-square test of equal population probabilities. Output displayed by the base R print() function, implicitly called by listing only the name of the component, contains line numbers and quotes around each line. The lessR function print_out() yields cleaner output.

This report also refers to two list components of statistics in Lines 15 and 26, b$n.miss, and b$p_value, called inline code references that begin with `r and end with `. Generating the report evaluates these R expressions with the resulting values embedded directly into the text of the report.

To generate the report, open a new R Markdown file from within RStudio from the New File option from the File menu. Copy the contents of the file into the window pane. Click the Knit button in the toolbar at the top of the window, which generates a web page complete with the bar chart, statistics, and commentary. To direct to a pdf file (if LaTeX is present) or an MS Word file (if Word or similar such as LibreOffice Writer is present), choose the arrow next to the Knit button, then select the appropriate alternative.

3.6   Part-Whole Visualizations

To compare the numerical values across the different levels of a categorical variable, bar charts and scatterplots display the value of a numerical variable for each level on a linear scale. These visualizations, however, do not present each numerical value as part of the whole set of values. The visualizations display the pieces, but not the whole. The pie chart, waffle plot, and treemap provide the perspective of the whole divided into its component pieces.

3.6.1 Doughnut and Pie Charts

The pie chart divides a circle into slices, one slice for each level of the categorical variable. The area of each slice reflects the corresponding value of the numerical variable. The `ggplot2` visualization in Figure 3.20b is of the traditional pie chart for the count of employees in each of the five departments from the employee data set. A more modern version of the pie chart, in the form of what is called either the *doughnut chart* or *ring chart*, is shown from `lessR` in Figure 3.20a. The doughnut chart replaces the center of a pie chart with a hole.

pie chart: A circle divided into sections, each sized according to the numerical value for the corresponding category.

employee data table, Section 1.2.2, p. 7

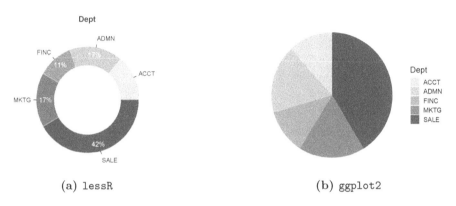

(a) `lessR`   (b) `ggplot2`

Figure 3.20: Doughnut (ring) chart and traditional pie chart.

The classical pie chart, however, is typically not recommended. Summarizing work by an influential contributor to data visualizations, Cleveland (1993), the R help manual for the base R pie chart function, `pie()`, bluntly informs: "Pie charts are a very bad way of displaying information. The eye is good at judging linear measures and bad at judging relative areas." People more accurately compare the length of lines (or bars) than the areas of slices of a circle. Particularly for numerical values almost equal to each other, differentiation according to length is more accurate than area.

Another attribute by which to compare the pieces of the pie is the angle of each piece that emanates from the center. As with areas, people do not proficiently compare angles as lengths. One more attribute for comparison is the circumference of each slice. Figure 3.21 illustrates the three characteristics for which to assess the size of pie slice: area, angle, and circumference.

pie chart dimensions: Area, angle, and slice.

In practice, outside of the academic/scientific literature, however, the pie chart is frequently encountered and offers advantages (Spence & Lewandowsky, 1991), probably more often in the form of the more modern doughnut chart. Without the center of the pie, perhaps viewers more accurately compare the size of the pie slices according to the circumference of the

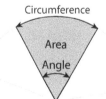

Figure 3.21: Three criteria to assess the size of a pie slice.

corresponding slices, analogous to the length of bars. Comparing the length of the edge of each slice at least to some extent mitigates the problem with comparing areas and angles.

Accordingly, the `lessR` function `PieChart()` by default creates a doughnut chart, with the `hole` parameter set at 0.65. Larger values up to but less than 1.0 generate a larger hole, that is, smaller ring. Smaller values result in a smaller hole. The value of 0 results in a classic pie chart with no hole. As with the bar chart, by default display a numerical value for each slice according to the `values` parameter, which defaults to `"%"` for tabulated counts and `"input"` for entered values of any numerical value y read from the data. Set `values` to `"off"` to not display the values.

values parameter,
Section 3.3.4, p. 60

`ggplot2` presents the pie chart as a stacked bar chart plotted with polar coordinates. Create a stacked one-column bar chart, then convert to polar coordinates with the `coord_polar()` function. Specify no x variable, indicated by an empty character string for the x-variable name. The y-variable is the numerical variable on which the size of each slice depends. Here y is not given but instead tabulated as the frequency of occurrence for each level of the categorical variable of interest, *Dept*, indicated by `stat(count)`. Map the levels of *Dept* to the `fill` color. The `coord_polar()` function adds some other details to the plot, best deleted for clarity, accomplished with the call to `theme_void()`.

R Input *Default doughnut/pie chart of Counts*

data: `d <- Read("http://lessRstats.com/data/employee.csv")`

---

lessR: `PieChart(Dept)`

ggplot2:
```
ggplot(na.omit(d), aes(x="", y=stat(count), fill=Dept)) +
    geom_bar() + coord_polar(theta="y") +
    scale_fill_grey(start=.8, end=.3) +
    theme_void()
```

---

qualitative scale,
Section 10.2.1, p. 214

Neither of the doughnut/pie charts in Figure 3.20 default to the displayed grayscale. Instead, both employ their respective default qualitative scale composed of multiple hues. Unless the levels of the categorical variable are ordered by magnitude, the qualitative scale is generally more appropriate than the sequential scale, the grayscale, presented here. With grayscale printing, however, either identically color each slice of the pie the same shade of gray, or present a sequential grayscale.

*pre-defined sequential
scales*, Section 10.2.2,
p. 218

With `lessR`, create the grayscale for all elements of the plot and subsequent plots with `style("gray")`. Or, set the `fill` parameter to `"grays"`. For `ggplot2`, invoke `scale_fill_grey()`. Here set the `start` and `end` points of 0.8 and 0.3 to somewhat compress the range of lightness/darkness for the displayed grayscale.

waffle plot: Colored
squares indicate the
value of the
numerical variable
for each category.

3.6.2   The Waffle Plot

The *waffle plot*, or square pie chart, replaces the slices of a pie with squares (of a waffle). For larger numerical values associated with a category, display more squares, as illustrated in Figure 3.22 for the variable *Dept* from the employee data set.

employee data table,
Section 1.2.2, p. 7

Create the waffle plot in Figure 3.22 with the `waffle()` function from Bob Rudis and Dave Gandy's `waffle` package (2017). Easily construct the waffle plot with the

Figure 3.22: Waffle plot of the counts for the five categories of Dept.

squares for each level displayed with a different hue. The `waffle()` function does not tabulate the frequencies of occurrence for each level of the categorical variable, so if the numerical variable y contains the counts, then separately compute such as with the base R `table` function. In this example, save the frequency table as myFreq. Obtain the corresponding waffle chart from `waffle(myFreq)`.

R Input *Waffle plot of counts*

data: `d <- Read("http://lessRstats.com/data/employee.csv")`
     `myFreq <- table(d$Dept)`

---

gray scale: `clr <- getColors("grays", n=5)`
waffle: `waffle(myFreq, colors=clr, flip=TRUE, xlab="Employees by Dept")`

Pass customized colors to `waffle()` with the `colors` parameter, one color for each category. To generate the grayscale visualization in Figure 3.22, first generate the sequential palette of five grayscale colors, here from the `lessR` function `getColors()`. Of course, any other of the `lessR` pre-defined sequential scales such as `"blues"` could be used in place of `"grays"`, or `"colors"` to replace the default `waffle()` qualitative scale with somewhat darker colors than the `waffle` default.

getColors(), lessR,
Section 10.2.1, p. 215

By default, the waffle chart displays vertically. Set `flip` to `TRUE` to orient horizontally. The standard base R parameter `xlab` provides the label.

3.6.3   The Treemap

The pie chart, waffle chart, and treemap all visualize the relations of individual parts, the proportions, to the whole. Unlike the pie chart, which works best with a relatively small number of slices, the treemap potentially displays many more elements. Moreover, the categories can be hierarchical, sub-categories within categories. For example, organize public schools by school district. Visualize learning success for a subject at a specific grade-level for the different schools displayed within their districts. The resulting tree map permits a comparison of both the schools and the districts. As another example, organize different products sold in a general-purpose store into different product categories such as hardware, clothing, and furniture. Then divide each product category into specific products. Visualize sales of the different products in different categories.

Trees – real trees and treemaps – have branches. Each branch of a treemap represents a category that displays as a rectangle, and each rectangle can further subdivide into component rectangles to represent the sub-categories. The treemap, as with

treemap: A
rectangle divided
into smaller
rectangles and
possible
sub-rectangles.

the pie chart and waffle plot, displays proportions of a whole by varying the size of the corresponding shape, here rectangles.

Martijn Tennekes' `treemap()` package (2017) provides a function of the same name to visualize treemaps. The first required parameter to the function is the data frame that contains the relevant data. The second required parameter, `index`, identifies one or more categorical variables, either a single variable or, for a hierarchy, a vector of categorical variables with the highest variable in the hierarchy listed first. The third required parameter, `vSize`, specifies the continuous variable that sets the scaling of the rectangle sizes according to the sum of its value for each category. The fourth parameter, required for some analyses, `vColor`, specifies the rectangle colors, in conjunction with the `type` of analysis. The parameters `vSize` and `vColor` may be the same variable.

One-variable treemap. The treemap in Figure 3.23 corresponds to the pie chart in Figure 3.20 and the waffle chart in Figure 3.22. The treemap illustrates the number of employees in each of the five company departments. The darkest and largest rectangle represents the department with the most employees, sales, with 15 employees. The lightest and smallest rectangle represents the department with the fewest employees, finance with four employees.

qualitative scale,
Section 10.2.1, p. 214

Although shown in grayscale, this treemap typically displays with a qualitative scale, each rectangle colored with a different hue, the default. The shades of gray in Figure 3.23 are alphabetical: the accounting department has the lightest gray and the sales department the darkest gray. There is no underlying scaling illustrated by a progression of light to dark colors consistent with a sequential scale.

sequential scale,
Section 10.2.2, p. 218

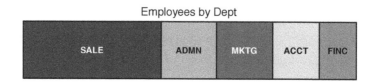

Figure 3.23: Treemap of the counts for the five categories of Dept.

Data input into the `treemap()` function may be the original data table, or a table of summary statistics computed from the full data table. In this first example, the treemap displays the number of employees in each department. Explicitly obtain these counts with the tidyverse `dplyr` functions `group_by()` to define the groups, corresponding to the five levels of *Dept*, then `summarise()` to calculate the corresponding counts, the created variable *Freq*. Store the summary table of *Dept* and *Freq* in myFreq, then input into `treemap()`. The summary table has six rows, one for each department, and one for missing data, which `treemap()` does not display.

> **R Input** *Treemap of counts*
>
> *data*: d <- Read("http://lessRstats.com/data/employee.csv")
>   d %>% group_by(Dept) %>% summarise(Freq=n()) -> myFreq
>
> *treemap*: treemap(myFreq, "Dept", "Freq", type="index",
>   title="Employees by Dept", palette=getColors("grays", n=5))

Setting `type` to `"index"` specifies to color the rectangles according to the value of the categorical variables that generate the classifications, here only one categorical variable, *Dept*. The `palette` parameter specifies the colors that fill the rectangles. In this example generate the grayscale with the `lessR` function `getColors()`.

getColors function,
Section 10.2.1, p. 214

Color that indicates magnitude. Color can also portray the magnitude of a continuous variable according to the classification of one or more categorical variables. The treemap in Figure 3.24 illustrates the distribution of mean *Salary* for each of the five departments. The color of each rectangle depends on the magnitude of mean *Salary*. The department with the largest mean *Salary*, $71,277, administration, has the darkest rectangle, and the department with the smallest mean *Salary*, $51,793, has the lightest colored rectangle. The size of each rectangle depends on the sum of the salaries for the corresponding department. Here the sales department has the most employees, and so the largest sum of salaries.

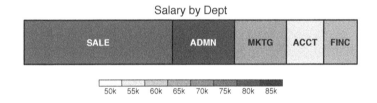

Figure 3.24: Treemap of mean salary for the five categories of Dept.

By default, `treemap()` colors the rectangles according to the sum of the relevant continuous variable(s). To analyze means, invoke the `fun.aggregate` parameter to specify the `"mean"` function. Set `type` to `"value"` to have the rectangle colors reflect the size of the corresponding means. The default legend title is not meaningful in this context, so remove it.

> **R Input** *Treemap of mean* Salary *for the five categories of Dept*
>
> *data*: d <- Read("http://lessRstats.com/data/employee.csv")
>
> *treemap*: treemap(d, "Dept", "Salary", vColor="Salary", type="value",
>   fun.aggregate="mean", title.legend="", title="Salary by Dept",
>   palette=getColors("grays", n=8))

Again, override the default color palette with grayscale to create Figure 3.24. No `palette` parameter specified results in the default colors, an orange-red palette that begins with a light orange for the smallest value and progresses sequentially until reaching a dark red for the largest value. A palette with eight colors exbibits a relatively wide range of luminance. Smaller values of n for getColors() provide progressively smaller ranges of luminance.

Hierarchical treemap. To obtain a hierarchical display, provide more than a single classification variable, what treemap() refers to as an index variable. From a 1993 data set (Lock, 1993), Figure 3.25 illustrates highway miles per gallon fuel usage for different types of automobiles, further classified by those made in the USA and those made elsewhere. The darkest square, for Small cars not made in the USA (0), indicates the cars with the best fuel mileage, 37.3 MPG. The two darkest squares belong to the Small car category, so the second best fuel mileage is for Small cars made in the USA (1), 34.3 MPG. The lightest colored squares are for Vans, corresponding to the worst fuel mileage, 22.6 MPG for non-USA Vans (0), and 21.5 MPG for USA Vans (0), the worst fuel mileage of all the categories.

In this example, only the Large type of car was manufactured in the USA, no instances of Large non-USA cars exist in these data. Accordingly, there is no subdivision of the Large rectangle into sub-categories, though treemap() does not label the sub-category that is present.

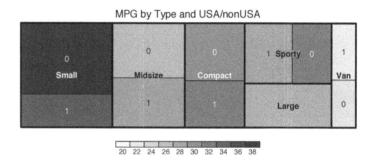

Figure 3.25: Hierarchical treemap of MPG.

To produce this treemap set vSize to c("Type", "Source"), which specifies *Type* as the first level of classification, with *Source* at the second level. As with the previous example, color the rectangles with the mean for each classification, here *MPG*. The size of the rectangles follows from the total *MPG* for each group.

R Input *Hierarchical treemap of the counts for Gender within Dept*

data: d <- Read("http://lessRstats.com/data/Cars93.csv")
       d %>% group_by(Dept, Gender) %>% summarise(Freq=n()) -> myFreq

---

treemap: treemap(d, c("Type", "Source"), vSize="MPGhiway",
        vColor="MPGhiway", type="dens", fun.aggregate="mean",
        title.legend="", title="MPG by Type and USA/nonUSA",
        palette=getColors("grays", n=8))

The treemap() function offers several different types of coloring methods of the rectangles, with "index" and "values" previously illustrated. In this example, "values" also applies, but the resulting relatively small range of MPG's resulted in a restricted range of luminance for the various rectangles. To provide for a larger range of variation, base the colors on the densities of MPG, specified by setting `type` to `"dens"`.

The `treemap()` function provides the computed values the treemap displays. To obtain these values, assign the function output to an object with any valid R name, such as `tm` in this example.

```
treemap: tm <- treemap(parameter values)
         tm
```

To display the values, enter the name of the object at the R console. The output includes a row for each combination of values of the levels of the categorical variables, the corresponding values of `vSize` and `vColor`, the coordinates and dimensions of the rectangles, and the color of each rectangle.

Chapter 4

Visualize a Continuous Variable

histogram, Figure 2.5,
p. 36

VBS plot,
Section 4.5, p. 94

A visualization of the distribution of a continuous variable estimates the shape of the underlying population distribution from a sample of that population. Chapter 2 introduced two such visualizations: The histogram as well as a modern replacement, the VBS plot. This chapter expands upon this material for both.

4.1   Histogram

4.1.1   Binning Continuous Data

Group measurements of a continuous variable into bins. In contrast to the relatively few unique values of a categorical value, a continuous variable has too many unique data values to plot individually. Consider annual USD salary for employees of a company. The salaries are in the ranges of tens of thousands of dollars, but each penny must be considered. Many, if not most, potential values never occur in the data, or once or a few times at most. For example, a *Salary* of $69,424.79 would likely not occur unless there is a huge sample size.

What to do with the unique data values that cannot meaningfully be individually plotted? There are many answers, some based on modern computer technology, which arguably yield a more satisfactory result, such as the VBS plot, than do older techniques. The traditional answer, based on 19th-century paper and pencil technology, partitions the range of values into *bins*, sometimes called *classes*. Each bin contains similar data values. Figure 4.1 presents an example of a bin that contains values from $60,000 to $70,000 for annual USD salaries.

bins: Sequence of
adjacent intervals,
each generally of the
same size.

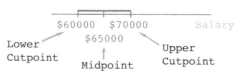

Figure 4.1: Example of a bin defined over the range of data values from $60,000 up to $70,000.

cutpoints: Lower,
upper boundaries of
a bin.
bin width: Width
of a bin.

midpoint: Value in
the middle of a bin,
summary of all
values in the bin.

Define each bin by its lower and upper boundary, its *cutpoints*, which sets the width of each bin, the *bin width*. The single point that most effectively summarizes the values of the bin is the *midpoint*, which is not necessarily a data value. In Figure 4.1, bin width is $10,000 and the midpoint is $65,000. Place each data value into its respective bin. Consistently assign values precisely equal to a cutpoint to either the adjacent lower bin or the adjacent higher bin.

histogram, Figure 2.5,
p. 36

histogram: Place
each data value of a
continuous variable
into its bin, plotted
as a bar with height
proportional to its
frequency.

The histogram. The most common visualization of a continuous variable is the *histogram*. To create the histogram, bin the data, that is, sort the individual data values into bins. Like the bar graph, the histogram consists of bars. Unlike the bar graph, the adjacent bars of a histogram share a common side to indicate the underlying continuity of a continuous variable.

The default `lessR` histogram, and a `ggplot2` histogram, both from Figure 2.5, are repeated here in Figure 4.2 for continuity. By default, `ggplot2` plots a histogram with a bin width that is deliberately too small to be meaningful, presumably to encourage exploration of different bin widths. Both histograms in Figure 2.5 have a bin width of $10,000 but differ because of different starting points.

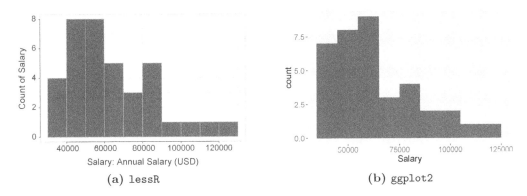

| |
| (a) `lessR`   (b) `ggplot2` |

Figure 4.2: Default histogram.

The `lessR` function `Histogram()`, abbreviated `hs()`, implements the base R histogram function, `hist()`. `Histogram()` retains the `hist()` defaults such as for computing bin width, as well as direct access to many of its parameters. By default, R assigns a value that is equal to a cutpoint to the lower bin, a rule also retained by `Histogram()`. As shown in Listing 2.2, `Histogram()` generates a statistical analysis in addition to the visualization, which includes the frequency distribution table, summary statistics, and an outlier analysis.

Histogram() output, Listing 2.2, p. 37

Obtain the `ggplot2` histogram with `geom_histogram()`. Customize bin width with the `binwidth` parameter.

> **R Input** *Default histogram*
> *data*: `d <- Read("http://lessRstats.com/data/employee.csv")`
> ------
> *lessR*: `Histogram(Salary)`
> *ggplot2*: `ggplot(d, aes(Salary)) + geom_histogram(binwidth=10000)`

fill and color parameters, Section 10.1.2, p. 213

sequential color palettes, Section 10.2.1, p. 216

The standard `fill` and `color` parameters apply to the creation of histograms. In addition to a single color, color palettes, such as the `lessR` sequential palette `"blues"`, are also available, with the potential for custom values of `chroma` and `luminance`. Or, customize individual bars. In this example with ten bars, the first four bars are blue, then one red bar, followed by five more blue ones.

> `Histogram(Salary, fill=c(rep("blue",4), "red", rep("blue",5)))`

To customize the fill colors, sequentially specify each bin fill color. The base R repetition function, `rep()`, avoids repetitively typing the colors, such as `"blue"` five times.

4.1.2   Histogram Artifacts

For the data values of any continuous variable, many different histograms can be created. Each histogram's shape depends on both the starting point of the bins as well as the bin width. Change one or both of these values, and the shape of the histogram may change, as illustrated in Figure 4.2. For example, the histograms from `lessR` and `ggplot2` are of the same data with the same bin width of $10,000, yet they differ. Why? The `lessR` histogram, following the computations of the base R function `hist()`, for these data values has a default start point for the first bin at $30,000. Set the `bin_start` parameter to $35,000 results in the same histogram computed by `ggplot2` in Figure 4.2b.

What are the correct bin width and the correct bin starting point for the data? The answer: none. In general, choose the smallest reasonable bin size supported by the available data to display the underlying distribution. However, the algorithms that choose a default bin width do not automatically provide the narrowest useful bin size. The selected value also depends on personal preference.

Bin Shift artifact: Shape of the histogram depends on the bin starting point

By trial and error, given the default histogram, define a more representative visualization of the corresponding population distribution. Usually the underlying population distribution is, or at least resembles, a smooth curve in one direction, or, as with the normal curve, a smooth curve to a maximum (or, inverted, minimum), and then a smooth curve in the other direction.

Under-smoothing: Bin width too small relative to the available data so that too many bins result in too much detail.

Consider the histograms in Figure 4.3, with very narrow or very wide bins.

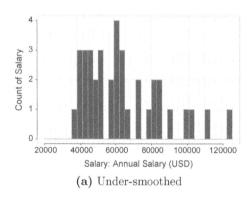
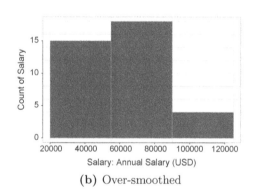

(a) Under-smoothed   (b) Over-smoothed

Figure 4.3: Under-smoothed and over-smoothed histograms that need adjustment of bin width.

The narrow binned histogram results from the `ggplot2` bin width default of approximately $3500, which results in the histogram in Figure 4.3a. The large number of ups and downs of the histogram bars more likely indicate too small of a bin width for the given sample size instead of a reflection of the actual population distribution. These excessive ups and downs indicate *under-smoothing*. The many fluctuations likely represent sampling error and so would not replicate with a new sample of the same size.

The other extreme is *over-smoothing*, illustrated in Figure 4.3b. An over-smoothed histogram results in bins too wide given the sample size, obscuring some of the available information. Instead, narrower bins result in more detail for the portrayal of the underlying distribution.

Over-smoothing: Not enough bins results in the bin width too large, obscuring properties of the underlying distribution.

The function calls to generate the histograms in Figure 4.3 appear below, which results from changing the value of `bin_width` in the call to the `lessR` function `Histogram()`.

> *under-smoothed*: `Histogram(Salary, bin_width=3000)`
> *over-smoothed*: `Histogram(Salary, bin_width=35000)`

No one value necessarily defines the best histogram, but the histogram in Figure 4.4, with a bin width of $12,000, represents a better histogram for these employment data than those previously presented.

To obtain a more representative histogram may require adjusting both the bin widths and the starting point of the bins, such as with the `lessR bin_start` parameter. In this example, `bin_width` is $18,000 and `bin_start` is $23,000.

4.1.3 Cumulative Histogram

Compute the cumulative histograms from the cumulative frequencies. For each bin, sum the counts from the first bin to that bin. Figure 4.5 shows an example.

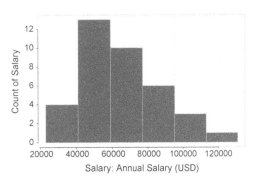

Figure 4.4: A more optimal histogram.

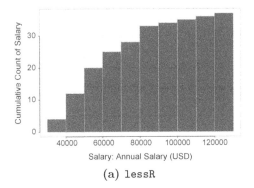

(a) `lessR`

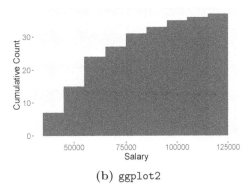

(b) `ggplot2`

Figure 4.5: Cumulative histogram.

For `lessR`, set the value for the parameter `cumulative` to `"on"` to obtain the cumulative histogram. For `ggplot2`, explicitly compute the cumulative frequencies with the base R function `cumsum()`, applied to the counts. Because the counts are not from the data, but rather computed by `ggplot2`, reference them as the y variable with the `stat()` function.

Also illustrated in this example is the flexibility of `ggplot2` in constructing multiple layers. Any `ggplot()` object can be saved, as in this example to the object `p`, displayed as is, and then another layer added for a subsequent visualization. To display the visualization `p`, as with any printable R object, enter its name at the console.

R Input

data: d <- Read("http://lessRstats.com/data/employee.csv")

---

lessR: Histogram(Salary, cumulative="on")

ggplot2: p <- ggplot(d, aes(Salary)) +
          geom_histogram(aes(y=cumsum(stat(count))), binwidth=10000) +
          labs(y="Cumulative Count")
        p

Figure 4.6 illustrates further development of the previous cumulative histogram. Here superimpose the regular histogram over the cumulative histogram.

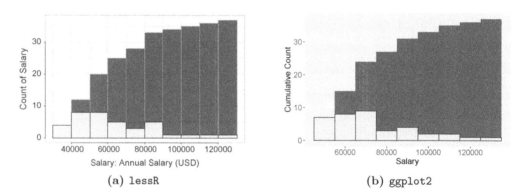

(a) `lessR`   (b) `ggplot2`

Figure 4.6: Cumulative histogram with a superimposed regular histogram.

For `lessR`, set the value of `cumulative` to `"both"` to generate the two layers, the regular histogram superimposed on the cumulative version. For `ggplot2`, explicitly add another histogram layer computed from the given data unmodified. From the saved object `p` from the last example, here add the layer to `p`.

R Input

data: d <- Read("http://lessRstats.com/data/employee.csv")

---

lessR: Histogram(Salary, cumulative="both")

ggplot2: p + geom_histogram(binwidth=10000, fill="gray90",
                          color="gray20")

Superimposing the regular histogram over the cumulative version illustrates the relationship between the two histograms: The value of each cumulative histogram bar increases from the size of the previous bin.

4.1.4 Frequency Polygon

The frequency polygon replaces the bars of a histogram with the midpoints of the top of each bar connected by adjoining line segments, as shown in Figure 4.7.

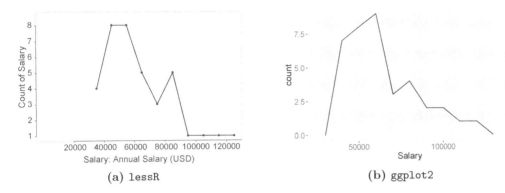

(a) `lessR`   (b) `ggplot2`

Figure 4.7: Default frequency polygons.

The frequency polygon is a scatterplot of each bin midpoint with the corresponding frequency, plus the connecting line segments. As such, the `lessR` function that plots scatterplots, `Plot()` or the abbreviation `sp()`, creates the frequency polygon. To indicate to plot the tabulated data and not the data itself, for the parameter `stat` set to the value `"count"`. For `ggplot2`, invoke `geom_freqpoly()`.

R Input *Frequency polygon*
data: d <- Read("http://lessRstats.com/data/employee.csv")

lessR: Plot(Salary, stat="count")
ggplot2: ggplot(d, aes(Salary)) + geom_freqpoly(binwidth=10000)

An advantage of the frequency polygon compared to the histogram is that the shape of the underlying distribution perhaps can be more clearly delineated. Still, the histogram and frequency polygon base the estimation of the likely smooth distribution on the approximation of bins. Perhaps best to provide the estimated smooth curve directly.

4.2 Density Plot

Histograms and frequency polygons were notable 19th-century innovations. More recently developed algorithms dependent on computer technology (Silverman, 1986; Sheather & Jones, 1991) directly estimate the smooth curve, the density curve, that likely underlies the distribution of data sampled for a continuous variable. As with the histogram, the smooth density curve indicates where the values of the variable tend to occur more or less frequently than the other values.

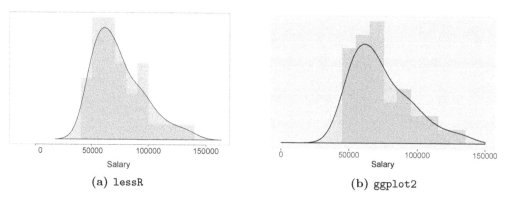

(a) `lessR`   (b) `ggplot2`

Figure 4.8: General and normal density plots with histogram.

4.2.1   Enhanced Density Plot

Figure 4.8 illustrates an enhanced density plot for `lessR` and `ggplot2`.

The `density` parameter for the `lessR` function `Histogram()` provides the general density curve, displayed over a histogram of the same data. With `ggplot2`, explicitly construct the visualization layer-by-layer, the histogram and density plot.

```
R Input  Density plots with histogram
data: d <- Read("http://lessRstats.com/data/employee.csv")

lessR: Histogram(Salary, density=TRUE)
ggplot2: ggplot(d, aes(Salary)) + xlim(0,150000) +
         geom_histogram(binwidth=10000, aes(y=stat(density)),
           color="gray80", fill="gray80") +
         geom_density(alpha=.4, fill="gray") +
         theme(axis.title.y=element_blank()) +
         theme(axis.text.y=element_blank()) +
         theme(axis.ticks.y=element_blank())
```

To obtain a normal density function, for `lessR` set the `type` parameter to `"normal"`, and to `"both"` to obtain both the general density curve and the normal curve. There is no `ggplot2` normal curve geom, so with the `ggplot2` `stat_function()` directly access the base R function `dnorm()` to plot normal densities.

```
R Input  Normal density plots with histogram
data: d <- Read("http://lessRstats.com/data/employee.csv")

lessR: Histogram(Salary, density=TRUE, type="normal")
ggplot2: add the following layer to the previous code:
         stat_function(fun=function(x)
           dnorm(x, mean=mean(d$Salary), sd=sd(d$Salary)))
```

To not include one of the three layers in Figure 4.8, the two density plots or the histogram, for `ggplot2` just do not include the corresponding layer in the construction of the visualization. To turn off the `lessR` histogram, set the `color_hist` and

`fill_hist` parameters to "off". To customize the histogram shape, apply the
`Histogram()` parameters `bin_start` and `bin_width`.

To customize the colors, for `ggplot2` apply the standard `fill` and `color` parameters
within the function call for each layer. To customize the colors in `lessR`, use
`color_gen`, `color_nrm`, and `color_hist` for the corresponding plotted curves, and
`fill_gen`, `fill_nrm`, and `fill_hist` for the corresponding fill colors.

As with most of `lessR` visualizations, a statistical analysis accompanies the visualization. The statistics for the density curve include the sample size, number of
missing values, the density bandwidth for potential adjustment in later analyses, the
Shapiro-Wilk normality test, and a box plot analysis of outliers described more fully
in a following section. Listing 4.1 presents the analysis that accompanies Figure 4.8.

```
Sample Size:  37
Missing Values:  0

Density bandwidth for general curve: 9529.0447
For a smoother curve, increase bandwidth with option: bw

Null hypothesis is a normal population
Shapiro-Wilk normality test:  W = 0.9117,  p-value = 0.0063

(Box plot) Outliers: 1

Small      Large
-----      -----
           124419.2
```

Listing 4.1: `Density()` statistical analysis.

4.2.2   Overlapping Density Curves

A common statistical analysis, a standard topic of introductory statistics textbooks,
compares the population means of a continuous variable across two different groups.
For example, is there a difference in average *Salary* for Men and Women at a given
company? The classic answer follows from the *t*-test for the mean difference, which
provides both a *p*-value regarding the null hypothesis of no difference, as well as the
corresponding confidence interval of the mean difference.

With modern computer graphics the default comparison can include the overlapping
density curves of *Salary* for Men and Women in Figure 4.9. As shown in Figure 4.9a,
the `lessR` visualization also includes both the descriptive and inferential statistics.
Also provided is a visualization of the sample mean difference expressed in the units
of analysis, here USD, and also expressed as the standardized effect size (*smd*),
Cohen's *d* (J. Cohen, 1988).

For `lessR`, obtain the statistical analysis and the visualization with the `ttest()`
function with the standard R model notation: response variable, *Salary*, on the left,
a tilde ~ , followed by the predictor variable, *Gender*. For the comparison of the
mean difference, the predictor variable is a grouping variable with only two values,
here M and F.

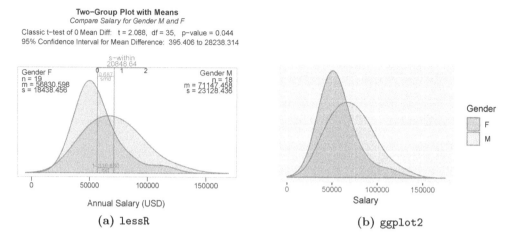

(a) <code>lessR</code>   (b) <code>ggplot2</code>

Figure 4.9: Overlapping density curves.

For `ggplot2`, to generate the overlapping density curves call `ggplot()` to access the data, the variables with `aes()`, and then the call to `geom_density()` with only the default parameter values. Within `aes()`, the first variable, *Salary*, is the variable for which to plot the densities. The `fill` parameter specifies the variable plotted separately for each group.

The optional `xlim()` function specifies the range of the values plotted over the *x*-axis. By default, `geom_density()` only plots over the range of obtained data, which usually, as in this situation, leaves the density curves truncated. Not extrapolating is the more conservative approach as the form of the data beyond the range of observation is speculative, but often a reasonable presumption. The call to the `theme()` function removes the `axis.title` of the word "density", and removes the density values and associated tick marks. The call to `scale_fill_manual()` sets grayscale for the fill colors of the two density curves.

R Input *Overlapping density curves*

data: d <- Read("http://lessRstats.com/data/employee.csv")

```
lessR: ttest(Salary ~ Gender)
ggplot2: ggplot(d, aes(Salary, fill=Gender)) + xlim(0, 175000) +
         geom_density(alpha=0.25, color="gray50", bw=14000) +
         theme(axis.title.y=element_blank(),
             axis.text.y=element_blank(),
             axis.ticks.y=element_blank()) +
         scale_fill_manual(values=c("gray25", "gray65"))
```

Obtain some of the `ggplot2` parameter values with trial-and-error. The default bandwidth was so low that the density curves had a wavy appearance, which likely reflects random noise. The limits of the *x*-axis needed for the full expression of the density curves are not known until after one or more iterations of modifying the density visualization.

The advantage of `ggplot2` here is that it is not limited to two density curves. However, as the number of groups increases, alternative plots such as box plots become more readable. An advantage of lessR is that a full statistical analysis accompanies the visualization. The first presentation is the descriptive statistics, shown in Listing 4.2.

```
Response Variable:   Salary, Annual Salary (USD)
Grouping Variable:   Gender, Male or Female

------ Description ------

Salary for Gender M:  n.miss = 0,   n = 18,   mean = 71147.458,   sd = 23128.436
Salary for Gender F:  n.miss = 0,   n = 19,   mean = 56830.598,   sd = 18438.456

Sample Mean Difference of Salary:   14316.860

Within-group Standard Deviation:    20848.636
```

Listing 4.2: ttest() descriptive statistics.

The next section of statistical output evaluates the assumptions of the t-test, shown in Listing 4.3.

```
------ Assumptions ------

Note: These hypothesis tests can perform poorly, and the
      t-test is typically robust to violations of assumptions.
      Use as heuristic guides instead of interpreting literally.

Null hypothesis, for each group, is a normal distribution of Salary.
Group M  Shapiro-Wilk normality test:  W = 0.962,  p-value = 0.647
Group F  Shapiro-Wilk normality test:  W = 0.828,  p-value = 0.003

Null hypothesis is equal variances of Salary, i.e., homogeneous.
Variance Ratio test:  F = 534924536.348/339976675.129 = 1.573,
    df = 17;18,  p-value = 0.349
Levene's test, Brown-Forsythe:  t = 1.302,  df = 35,  p-value = 0.201
```

Listing 4.3: ttest() evaluation of assumptions.

The inferential tests are conducted with and without the assumption of homogeneity of variance. The analysis with the assumption of homogeneity appears in Listing 4.4.

```
------ Inference ------

--- Assume equal population variances of Salary for each Gender

t-cutoff: tcut =  2.030
Standard Error of Mean Difference: SE =  6857.494

Hypothesis Test of 0 Mean Diff:  t = 2.088,  df = 35,  p-value = 0.044

Margin of Error for 95% Confidence Level:  13921.454
```

Listing 4.4: ttest() inferential analysis, assumption of homogeneity of variance.

Also provided are the effect size, and the bandwidth for each density curve, the latter of which may be adjusted with the bw parameter. There is also a version of the ttest() named tt_brief, which provides the same visualization, but only the basic statistical output of the summary statistics, hypothesis test, and confidence interval of the mean difference.

4.2.3   Rug Plot

rug plot: A type of scatterplot in which a small tick mark represents each individual data value.

Another version of a density plot is a *rug plot*, a scatterplot for a single variable that plots each data value as a tick mark instead of a dot. Typically display the rug along the variable axis of a density plot, as in Figure 4.10.

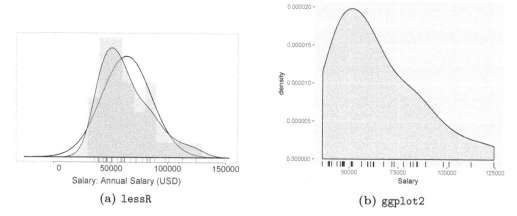

(a) `lessR`   (b) `ggplot2`

Figure 4.10: Density plots with rugs.

To create a rug plot, with `ggplot2` add the rug geom, `geom_rug()`. For `lessR`, set the `rug` parameter to `TRUE` in the call to `Histogram()`, also with `density` set to `TRUE`. To better compare the default `lessR` and `ggplot2` density plots, enhance the `ggplot2` visualization in Figure 4.10 with light gray fill and transparency for the plot of the density function.

R Input *Density plots with rug*

data: `d <- Read("http://lessRstats.com/data/employee.csv")`

---

lessR: `Histogram(Salary, density=TRUE, rug=TRUE)`

ggplot2: `ggplot(d, aes(Salary)) +`
          `geom_density(alpha=.4, fill="gray") + geom_rug()`

To change colors from the default black for the rug tick marks, add the `color` parameter to the function `geom_rug()`. For `lessR`, invoke the `color_rug` parameter. The optimal width of each tick mark in the rug depends on the number of data values. Here, use the `size` parameter for `ggplot2`, and the `size_rug` parameter for `lessR` to change the line widths of the tick marks.

violin plot: Mirrored density plot, usually longer than wider (or reverse).

4.2.4   Violin Plot

The density plot displays density with the vertical height from the origin. The related *violin plot* rotates the density plot to the mirror image of the densities, symmetric about its axis. See Figure 4.11 for an example. The violin plot displays density according to the width of the mirrored densities, which emphasizes more changes in density along the continuum than does a single density plot by itself.

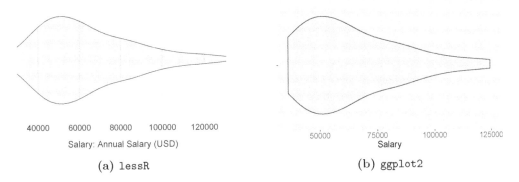

(a) `lessR`   (b) `ggplot2`

Figure 4.11: Violin plots.

To create the violin plots, with `ggplot2` add the `geom_violin()` layer. Here fill the violin with a light gray, partially transparent given the `alpha` setting of 0.4. The reference to `axis.title.y` removes the letter x that otherwise appears on the vertical axis. For `lessR`, invoke the `ViolinPlot()` function, or its abbreviation, `vp()`.

R Input *Violin plots*

data: `d <- Read("http://lessRstats.com/data/employee.csv")`

---

lessR: `ViolinPlot(Salary)`

ggplot2: `ggplot(d, aes(Salary)) +`
`        geom_violin(fill="gray85", alpha=0.4) + coord_flip() +`
`        theme(axis.title.y=element_blank())`

Both `lessR` function names, `ViolinPlot()` and `vp()`, are aliases to the `lessR` general `Plot()` function, with the parameter `vbs_plot` set to `"v"` for violin plot.

4.3   Box Plot

The boxplot (Tukey, 1977) follows from the *interquartile range (IQR)*. Sort the values of the distribution from smallest to largest. The *quartiles* are three values that divide the sorted distribution of values into quarters, four equal-size groups. The first one-quarter of the values lie between the smallest value and the first quartile. The second quartile is the median, the value that occupies the middle position between the smallest and largest values in the sorted distribution. The third quartile separates the largest 25% of the values from the smaller values. The IQR is the range of values between the 3rd and 1st quartiles.

The box in the box plot is drawn with the 1st and 3rd quartiles defining its (narrow) edges, which are then connected to form a box, as shown in Figure 4.12. The width of the box is irrelevant. This is the same box plot from the full VBS plot shown in Figure 2.6.

IQR: The interquartile range, the middle 50% of the values of the sorted distribution, the difference between the 3rd and 1st quartiles.

quartile: The three quartiles divide a sorted distribution into quarters.

box plot: A box around the 1st and 3rd quartiles, a line for the median, and lines (whiskers) out from the box to include all values within 1.5 IQR of each edge.

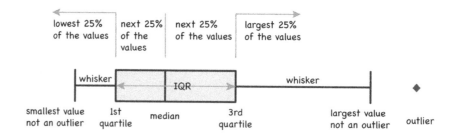

Figure 4.12: Box plot.

The box plot bins the distribution into quartiles. The distribution in Figure 4.12 skews to the right, indicated by the distance from the median to the right-and edge of the box.

4.3.1   Classic Box Plot

The standard box plot for the *Salary* data is shown in Figure 4.13.

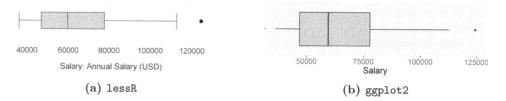

(a) `lessR`   (b) `ggplot2`

Figure 4.13: Box plot.

Obtain the `ggplot2` box plot with the `geom_boxplot()` layer. For `lessR`, use the `BoxPlot()` function, abbreviated `bx()`. By default, `ggplot2` displays a vertical box plot. Neither vertical nor horizontal is right or wrong, but a matter of preference. To show both box plots with the same orientation, flip the box plot horizontal with `coord_flip()`.

R Input *Box plot of* Salary *from* d *data frame*

data: d <- Read("http://lessRstats.com/data/employee.csv")

lessR: BoxPlot(Salary)

ggplot2: ggplot(d, aes(x="", y=Salary)) +
        geom_boxplot(fill="gray75") + coord_flip()

`BoxPlot()` and its abbreviation `bx()`, are both aliases to the `lessR` general `Plot()` function, with the parameter `vbs_plot` set to `"b"` for only the box plot.

A primary use of the box plot identifies potential outliers, necessary to understand the properties of a distribution (Belsley, Kuh, & Welsch, 1980; Osborne & Overbay, 2004). An outlier could indicate a data collection or transcription error. Or an outlier could be a data value sampled from a population distinct from the population that generated the remaining data values. A data value sampled from a different population would bias the analysis regarding generalizations to the intended population of interest. When an outlier is identified, the analyst should understand the process that generated the anomalous value.

Label values far from the edges of the box as outliers. For a symmetric distribution, define a potential outlier as a data value that lies 1.5 IQR's beyond the 1st or 3rd quartiles. Define an actual outlier as a value that lies 3.0 IQR's or more beyond these quartiles. By default, `lessR` typically plots both sets of outliers as individual points. Plot dark red for potential outliers and then a brighter red for actual outliers, unless grayscale is selected. Figure 4.13 identifies the highest salary as an outlier.

4.3.2 Box Plot Adjusted for Asymmetry

The IQR cutoffs of 1.5 and 3.0 more appropriately apply to symmetric distributions (Vandervieren & Hubert, 2004; Hubert & Vandervieren, 2008). Compared to a symmetric distribution, more values in the skewed tail of an asymmetric distribution are expected, perhaps even extreme values. Values in the opposite tail are expected to vary less compared to a symmetric distribution.

To illustrate, Figure 4.14a applies the classic boxplot definition of outliers to 250 randomly sampled values, all from the same asymmetric distribution, the standardized log-normal output from the base R `rlnorm()` function. The distribution is severely right-tail skewed, indicated by the 22 values classified as outliers according to the rule of more than 1.5 IQR's larger than the 3rd quartile. These values are outliers in the sense that they are far from most other values, yet they are expected given the characteristics of the underlying distribution.

(a) classic box plot   (b) adjusted box plot

Figure 4.14: Box plot and adjusted box plot for a right-tail skewed distribution.

To compensate, Hubert and Vandervieren (2008) generalize the classic 1.5 IQR's definition to also consider the distribution's skewness. Introduce two new parameters, a and b, whose values adjust the amount of change of the whiskers on both sides of the box. With non-zero values of a and b, the whiskers on both sides of the box plot are shifted toward the direction of skew. Without skew, the adjusted formula simplifies to the classic definition of an outlier. As shown in Figure 4.14b, for the adjusted box plot the right whisker becomes larger, reducing the number of detected outliers.

The R package `robustbase` (Maechler et al., 2016) provides for this box plot adjustment. The `lessR` function `Plot()`, and its aliases `BoxPlot()` and abbreviation `bx()`, can implicitly call functions from **robustbase**. Set the `Plot()` parameter `box.adj` to TRUE. Perhaps change the parameters `a` and `b` from their respective default values of -4 and 3 to change the impact of the adjustment of the whiskers on each side of the box.

4.4   One-Variable Scatterplot

jitter: Random perturbation of the plotted points in a scatterplot.

The scatterplot provides for the direct visualization of the data values and corresponding sample size. Each data value plots as a point, according to its location along the corresponding value axis. Points corresponding to the same value overlap. Plot overlapping points with random perturbations called *jitter* to distinguish between multiple points of the same value, shown in Figure 4.15.

(a) `lessR`   (b) `ggplot2`

Figure 4.15: Box plot.

lessR automatically applies jitter to the extent that overlapping occurs. More overlap results in more jitter, vertical jitter if possible to retain the value of the plotted point, but horizontal jitter is applied if there are too many points with the same value. For **ggplot2**, to obtain jitter use `geom_jitter` instead of `geom_point` to plot the points. To show both box plots with the same orientation, flip the box plot horizontal with `coord_flip()`.

R Input *Box plot of* Salary *from* d *data frame*

data: `d <- Read("http://lessRstats.com/data/employee.csv")`

lessR: `ScatterPlot(Salary)`

ggplot2: `ggplot(d, aes(x="", y=Salary)) +`
`        geom_jitter(position=position_jitter(0.02)) + coord_flip() +`
`        theme(axis.title.y=element_blank())`

As with the violin and box plots, `ScatterPlot()`, or its abbreviation `sp()`, are aliases for specific parameter values of the more general `Plot()` function. The equivalence is a function call to `Plot()` with `vbs_plot` set to `"s"`.

4.5   Integrated Violin/Box/Scatterplot

Histogram(), Chapter 4, p. 79

Since the 19th century the histogram, such as in Figure 2.5, has been the standard visualization to display the frequencies of a continuous variable such as *Age*, *Salary*,

MPG or *Height.* The problem is that the histogram follows from a simple technology developed well before the first computer was ever built and the first computer data visualization ever generated. Its uses jagged, discontinuous bars to represent an underlying continuity, does not detect outliers, does not provide a visualization of sample size, and does not easily stack to compare against different groups (Gerbing, 2020). No surprise that modern computer technology provides a means for which to develop more informative displays of the distribution of the values of a continuous variable than the histogram. We can do better with modern tools.

4.5.1 VBS Plot

The VBS plot (Gerbing, 2020) integrates three streamlined visualizations of the distribution of a continuous variable: violin, box, and scatter plots. The VBS plot of the *Salary* data was shown in Figure 2.6, but is repeated here in Figure 4.16 for continuity.

VBS plot, Figure 2.6, p. 37

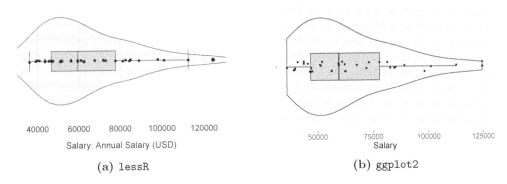

(a) `lessR`   (b) `ggplot2`

Figure 4.16: Integrated Violin/Box/Scatterplot or VBS plot.

For the `Plot()` function, override the defaults to remove individual components of the plot. To request fewer components, use only one or two of the letters `"v"`, `"b"` and `"s"` for the `vbs_plot` parameter, which has a default value of `"vbs"`, that is, by default yields the full integration.

For `ggplot2`, add any combination of the three layers with calls to the corresponding three `geom` functions to specify the three types of geometric objects to plot: `geom_violin()`, `geom_boxplot()`, and `geom_jitter()` functions according to the same general form as illustrated in Figure 2.2. The `geom_jitter()` function jitters and then plots the individual points of the scatterplot. The `coord_flip()` function call flips the orientation of the plot from vertical to horizontal, which is solely a user preference. The reference to `axis.title.y` removes a superfluous label on the vertical axis.

ggplot2 code, VBS plot, Chapter 2.3.2, p. 38

The `ggplot2` VBS plot, however, must be constructed over repeated analyses to determine the ideal bandwidth, the amount of smoothness in the violin plot, as indicated by the `bw` parameter. Adjust the width of the boxplot that fits inside of the violin plot with the `width` parameter. The `position` parameter sets the amount of jitter. `Plot()` performs these computations automatically to adjust, for example, the size of the points from sample sizes from 5 to 50,000. Also note from Figure 4.16 that `geom_jitter()` plots the outlier, as does `geom_boxplot()`, so the data input into `geom_jitter()` should be restricted to not plot this point twice.

`Plot()` also provides the statistical analysis in Listing 4.5 along with the visualization. These statistics include summary statistics including those of the box plot, labeling of the outliers, and the computed parameter values to construct the visualization.

```
--- Salary ---
Present: 37
Missing: 0
Total  : 37

Mean         : 63795.557
Stnd Dev     : 21799.533
IQR          : 31012.560
Skew         : 0.190   [medcouple, -1 to 1]

Minimum      : 36124.970
Lower Whisker: 36124.970
1st Quartile : 46772.950
Median       : 59547.600
3rd Quartile : 77785.510
Upper Whisker: 112563.380
Maximum      : 124419.230

(Box plot) Outliers: 1

Small       Large
-----       -----
            18 124419.23

Number of duplicated values: 0

Parameter values (can be manually set)
------------------------------------------------------
size: 0.61      size of plotted points
jitter.y: 0.45  random vertical movement of points
jitter.x: 0.00  random horizontal movement of points
bw: 9529.04     set bandwidth higher for smoother edges
```

Listing 4.5: VBS statistical analysis.

Knowing the values of these computed parameters is useful to implement customizations manually. For example, the smoothing algorithm sometimes tends toward monotonicity, so reduce a too smooth bandwidth by re-running with a lower value of `bw`.

The trade-off between `lessR` and `ggplot2` is clear. The simpler `lessR` functions produce pre-defined visualizations, which, for many situations provides the intended visualization, along with a corresponding statistical analysis. For example, with the VBS plot, all three components provide unique, but complementary information, in not much more space than is required for any one of the components. In contrast, although the `ggplot2` construction is more involved with multiple function calls, many `geom` functions are available that provide the analyst with an incredibly versatile tool kit for customizing visualizations.

4.5.2 VBS Plot of Likert Data

To further illustrate how `Plot()` adjusts the underlying parameters to the characteristics of a specific visualization, consider the VBS plot of the responses to an attitude item on a Likert scale, that is, usually about four to seven response possibilities such as Strongly Disagree, Disagree, etc. The example here, shown in Figure 4.17, is of the analysis of Machiavellianism with 351 responses (Hunter et al., 1982) to the original Mach IV scale (Christie & Geis, 1970). The underlying attitude varies along a continuum of disagreement/agreement, but is only assessed with six unique responses from Strongly Disagree (0) to Strongly Agree (5).

`Plot()` detects the large number repeated data values, just six values distributed over 351 responses. In addition to the previous example that included only vertical jitter, `Plot()` appropriately adds horizontal jitter, which moves the point off of its coordinate but fits the scatterplot within the enclosing violin plot. Yet given the few unique data values, the default `Plot()` densities recover the underlying continuity. The densities demonstrate a monotonic smooth decrease in agreement from the

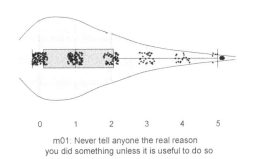

Mach IV scale, Section 1.2.6, p. 22

VBS plot of Salary, Figure 4.16, p. 95

Figure 4.17: VBS plot of 6-pt Likert scale responses to the first item on the Mach IV scale.

Strongly Disagree position, the mode, through the Strongly Agree position, which are outliers in relation to the entire distribution. Compared to the histogram, the VBS plot better resolves the dialectic tension of highly discrete measurements from an underlying continuous distribution. In this application, the histogram would be limited to only six bins, whereas the VBS plot approximates the continuity, identifies outliers, and provides a visual display of sample size and responses for each response category.

The code to generate this visualization follows.

```
R Input Create a set of factor variables with variable labels
data: d <- Read("http://lessRstats.com/data/Mach4.csv")
      l <- Read("http://lessRstats.com/data/Mach4_lbl.csv",
              var_labels=TRUE)

lessR: Plot(m01)
```

With the R basics now presented, we are ready to create data visualizations.

Unlike the previous example of the Likert responses to the Mach IV items that converted the integer data values to factors, here analyze the data as numbers. As such, there are no value labels to label each of the six responses, but invoke the `lessR` variable labels for the Mach IV items, which display on the statistical output and visualizations. With `Read()`, set `var_labels` to `TRUE` to read the labels into the `l` data frame, which consists of two columns: the variable name and then the variable label.

stacked bubble plot (BPFM), Figure 3.4, p. 50

variable labels, Section 1.2.5, p. 18

4.5.3   Trellis Plots or Facets

Following is a simple illustration of Trellis graphics: Generate a VBS plot separately for Men and Women. The variable that specifies the levels at which to create the visualization is called a *conditioning variable*. In Figure 4.18 condition the histogram of *Salary* on the two levels of *Gender* present in these data: Male and Female.

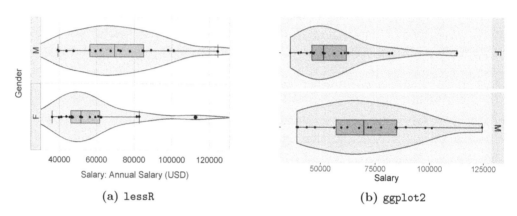

(a) `lessR`   (b) `ggplot2`

Figure 4.18: Trellis VBS plot, one panel each for Men and Women.

We conclude that a higher percentage of women occupy the lower salary ranges. The pattern of these visualization encourages further investigation to establish any significant differences in *Salary*, and perhaps a pattern of discrimination.

With `lessR`, Trellis visualizations follow from the `lattice` package (Sarkar, 2008), with most of `lessR` style and color parameters passed to the corresponding lattice functions. Specify the first conditioning variable with the `by1` parameter, and a second conditioning variable with `by2`. Another possibility yields a single panel graph, but with the scatterplots for the different levels of the categorical variable on the same plot.

same plot,
Section 5.2, p. 112

With `ggplot2`, invoke the `facet_grid()` function to specify Trellis graphics, referred to as facets within the `ggplot2` system. A feature of `ggplot2` is that it creates visualizations in increments by saving each cumulative increment into a `ggplot2` object. Store the previously provided instructions for the one panel VBS plot into object `p`, and then an additional specification to what was already developed to generate the multi-panel plot.

R Input *Define ggplot() object p for later use*
data: `d <- Read("http://lessRstats.com/data/employee.csv")`

```
p <- ggplot(d, aes(x="", y=Salary)) +
     geom_violin(fill="gray75", bw=9500) +
     geom_boxplot(fill="gray65", width=0.25) +
     geom_jitter(shape=16, position=position_jitter(0.01)) +
     coord_flip()
```

Here proceed with the simpler expression to produce a multi-panel display without repeating the work already accomplished. To display the created object, "print" it as any other object to display, enter the p at the R console, or follow with more layers separated by the + sign.

> **R Input** *VBS Plot of* Salary *from* d *data frame*
>
> *data:* d <- Read("http://lessRstats.com/data/employee.csv")
> ───
> *lessR:* Plot(Salary, by1=Gender)
> *ggplot2:* p <- p + facet_grid(cols=vars(Gender))
>   p

In the call to facet_grid(), set the cols parameter to vertically stack the two *Gender* VBS panels, which facilitates the comparison of the Salaries for men and women.

The focus here is a modern replacement for the pre-computer technology histogram. Plot a single VBS plot, or distribute the plots for levels of a categorical variable over one or more panels with lessR according to the following:

▷ Plot(X) for a VBS plot of variable X

▷ Plot(X, by=Y) for different groups of X according to Y plotted on the same panel

▷ Plot(X, by1=Y) for a Trellis VBS plot of variable X conditioned by categorical variable Y. For a second conditioned variable add the by2 parameter.

Further, according to the parameter vbs_plot, any combination of the violin plot, the box plot, and the scatterplot may be plotted. Choose any combination of "v", "b", and "s" as a single character string.

4.6   Pareto Chart

The *Pareto chart* consists of two separate plots of a categorical variable on the same panel. One plot, with values shown on the left vertical axis, is a bar chart of the counts of the categories of interest. The bars are sorted in descending order. The second plot, a frequency polygon, cumulates the same counts. The Pareto chart visualizes the most influential factors that contribute to some outcome.

bar chart,
Section 3.1.1, p. 47

cumulative distribution,
Section 4.1.3, p. 83

In 1906 the economist Vilfredo Pareto noted that about 20% of the Italian population owned 80% of the land, an observation that generalizes to a variety of types of wealth across many countries. To generalize further, a key observation in implementing quality control of a manufacturing process is that usually a small number of different types of defects lead to most of the quality issues. Generally, about 20% of defects lead to about 80% of the defective product. Or, about 20% of the reasons for returning merchandise leads to about 80% of the returns. On the positive side, about 80% of sales come from about 20% of customers. This statement that 80% of the effects originate from 20% of the causes is known as the *Pareto principle*.

The Pareto chart provides the visualization for this 80-20 rule, illustrated in Figure 4.19. The first Pareto chart, in Figure 4.19a is from the quality control chart package qcc. The second version in Figure 4.19b is from ggplot2.

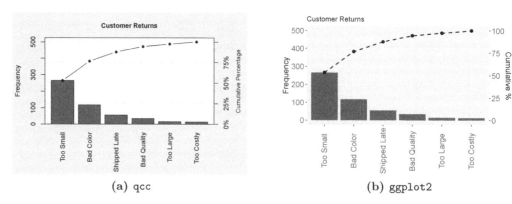

(a) `qcc`   (b) `ggplot2`

Figure 4.19: The Pareto chart.

Figure 4.19 reveals two primary reasons for the customer returning the purchased clothing: Too Small and Bad Color, which together account for 76.4% of the returns. From this analysis, the retailer understands that either the sizing of the clothing should be adjusted or a notice placed on the product pages on the website that sizing is a little smaller than many customers expect. Second, the photography on the website should be adjusted to more accurately display colors so that customers receive clothing with the colors that they expect.

The `qcc` package function `pareto.chart()` produces the Pareto chart with a simple function call. The only issue is that the frequencies for each category are entered as the data to the function call. Either enter the frequencies into a vector directly, or calculate them from base R `table()`. Here the data consists of a column of 500 rows of data, the data value for each row a reason for returning the clothing. Read into the data frame *d*. The first entry in this table, in Row 1, is Reason, the variable name. The function call `table(d$Reason)` yields the tabulations in Listing 4.6.

```
1  > myCount
2  Reason
3    Bad Color    Bad Quality   Shipped Late   Too Costly    Too Large    Too Small
4  quality control     34            56            13            15            265
```

Listing 4.6: Tabulated data input into the `qcc` Pareto chart function.

As shown in the following code, store the results of the tabulation in the vector *myCount*, which is then entered into `pareto.chart()`, along with the chart title, the standard base R `main` parameter, and the color here overriding the default blue scale, with the base R `col` parameter. The function does the work of sorting the tabulated values, calculating the cumulative sums, and plotting the separate layers.

R Input *qcc Pareto chart*

data: `d <- Read("http://lessRstats.com/data/Pareto.csv")`
      `myCount <- table(d$Reason)`

qcc: `pareto.chart(myCount, main="Customer Returns", col="gray30")`

To obtain the Pareto chart for `ggplot2` requires more work. There is a package `ggQC` that implements `ggplot2` functions to create a Pareto chart with the `stat_pareto()` function, but to understand more how `ggplot2` works, following is the R code to obtain the plot directly from `ggplot2`. The data preparation section is presented first, followed by the code for the Pareto chart itself. This example serves as a further example of the programming power of `ggplot2`, the potential to create a wide variety of visualizations from its basic function calls.

As with the function from `qcc`, begin by tabulating the counts for each of the reasons for returning the item of clothing, here converting the tabulated counts to a data frame named *d2*. The `table()` parameter `dnn`, dimension names, names the variable for the tabulated frequencies, which themselves are named *Freq*. Sort the tabulated values in decreasing order with the tidyverse `dplyr` function `arrange()`, with the variable *Freq* called from the `desc()` function to sort in descending order. This sorted ordered must be preserved for plotting, accomplished by ordering according to the existing ordering, codified as the levels of a factor. Next, compute the cumulated sums with base R `cumsum()`, storing them in the variable *Cml*.

ordered factor levels, Section 1.2.6, p. 20

R Input *Data preparation for ggplot2 Pareto chart*

```
data: d <- Read("http://lessRstats.com/data/Pareto.csv")
      myCount <- table(d$Reason, dnn="Reason")
      d2 <- data.frame(myCount)
      d2 <- arrange(d2, desc(Freq))
      d2$Reason <- factor(d2$Reason, levels=d2$Reason)
      d2$Cml <- cumsum(d2$Freq)
```

Data frame *d2* results from this data wrangling, which consists of the reasons for return and their associated counts and cumulated counts, shown in Listing 4.7. This is the data entered into the `ggplot()` function.

```
1  > d2
2          Reason Freq Cml
3  1    Too Small  265 265
4  2    Bad Color  117 382
5  3 Shipped Late   56 438
6  4  Bad Quality   34 472
7  5    Too Large   15 487
8  6   Too Costly   13 500
```

Listing 4.7: Data prepared to input into the `ggplot2` Pareto chart function calls.

To write the code to generate the `ggplot2` Pareto chart, first consider the component layers of the plot: a bar chart, a cumulative frequency distribution indicated by corresponding points on the plot, and a set of line segments to connect the points of the cumulative frequencies. Plot these three layers, respectively, with `geom_bar()` (or `geom_col()`), `geom_point()`, and `geom_line()` for the line that connects the points that plot the cumulative frequencies. Plot the bars with the y-coordinate of variable *Freq*. Plot the cumulative frequency points and corresponding line with the y-coordinates of the variable *Cml*.

The `geom_bar()` parameter `fill` sets the bar fill color. The `stat` parameter instructs to read the values of y from the data. For `geom_point()`, enlarge the points by setting `size` to 2.5. For `geom_line()`, the points must belong to the same group, but because *Reason* is a factor, each level of the factor represents a different group. To address this issue, define all data values in *d2* to belong to the same group by setting `group` to 1. Set the parameter `linetype` to plot a dashed line.

R Input *ggplot2 Pareto chart*

data: see previous for construction of data source d2

```
ggplot2: ggplot(d2, aes(Reason)) +
           geom_bar(aes(y=Freq), fill="gray30", stat="identity") +
           geom_point(aes(y=Cml), color="black", size=2.5) +
           geom_line(aes(y=Cml, group=1), color="black", size=.75,
             linetype="dashed") +
           theme(axis.text.x = element_text(angle=90, vjust=0.5)) +
             scale_y_continuous("Frequency",
               sec.axis=sec_axis(~ ./5, name="Cumulative %")) +
           labs(title="Customer Returns", x="", y="Frequency")
```

After plotting the three layers, further enhance the visualization with `theme()`. Specify `axis.text.x` to present the category values perpendicular to the x-axis. With `scale_y_continuous()`, create a transformed version of the vertical axis on the left with a parallel axis on the right. Here the transformation divides the values on the left-axis by 5, to convert to percentages as there are 500 rows of data in the original data frame *d2*. The `labs()` function adds a title, removes the x-axis label, and provides a y-axis label.

stat_pareto(),
ggQC package:
Function to construct
a Pareto chart.

ggQC control chart,
Figure 7.2, p. 161

The above `ggplot2` code provides a further example of `ggplot2` programming, including a second vertical axis. Another, more direct option, analogous to the `qcc` function in the `qcc` package, is the `stat_pareto()` function in the `ggQC` package, which, like `qcc`, provides a variety of quality control functions and visualizations. The `ggQC` control chart function is illustrated in a later chapter.

Chapter 5

Visualize the Relation of Two Continuous Variables

Relationship, continuous variables: As the values of one variable increase, the values of the other variables tend to either increase, or decrease.

How do two or more variables relate to each other? Chapter 6 presents analyses of the relationship between two categorical variables. Here focus on the *relationship* between continuous variables: As the values of one variable increase, the values of the other variables tend to either systematically increase (+ relationship) or systematically decrease (- relationship). Examples follow.

Positive Relationship:

  ○ Food quality increases, customer satisfaction increases

  ○ Hotel occupancy rate increases, needed staff increases

Negative (inverse) relationship:

  ○ Price decreases, sales volume increases

  ○ Time partying increases, grades decrease

Correlation coefficient: Indicates extent of a linear relationship of two variables, bounded by -1 and 1.

Positive and negative relationships can have the same magnitude, such as for linear relationships, assessed by the *correlation coefficient*. The sign of + or − indicates the direction of the relationship. The size of the coefficient indicates magnitude.

Scatterplot: One plotted point for each pair of values for two variables.

Quick Start: Scatterplot, Chapter 2.4, p. 38

The essential visualization for the relationship between two continuous variables is the *scatterplot*. The paired data values for each observation plot as a single point. Define the two coordinates of each point as the values of both variables for the corresponding observation.

5.1   Enhance the Scatterplot

5.1.1   The Ellipse

Bivariate normal distribution: 3-D analog of the 2-D normal curve.

A distribution of two variables, x and y, necessarily involves a third variable. For categorical variables, the third variable is their joint frequency at specific values of x and y. For continuous variables, the third variable is their joint density, z, which plots as a smooth curve. The plot extends into three dimensions with z represented as height. Consider the plot of a bivariate normal distribution, such as the two examples in Figure 5.1.

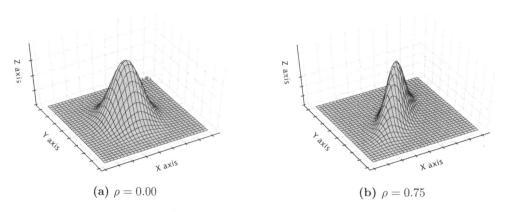

(a) $\rho = 0.00$ (b) $\rho = 0.75$

Figure 5.1: Bivariate normal distributions.

The plot represents a three-dimensional generalization of the two-dimensional normal curve. The fullness of the resulting 3-D object depends on the size of the correlation between x and y, here expressed in terms of their population correlation, ρ.

Summarize the relationship of two bivariate distributed variables with an ellipse. An ellipse results from a slice through this three-dimensional bivariate normal distribution parallel to the "floor", the region defined by the x- and y- axes. The ellipse is the edge of the slice, a contour curve with all points of equal height on the three-dimensional object. For a population correlation of $\rho = 0.00$, the ellipse is a circle if the variables are expressed in the same scale and same physical units. The larger the correlation between x and y, $+$ or $-$, the narrower the ellipse.

The size of the ellipse depends on the height of the slice through the object on the z-axis. The closer to the base, the x-y coordinate plane, the larger the ellipse, so the larger the confidence level. The usual choice for the confidence level is 0.95. That means that 95% of the data values from the corresponding bivariate normal distribution lie within the 0.95 confidence ellipse. For any one sample from a bivariate normal distribution, *approximately* 95% of the values fall within 0.95 confidence the ellipse.

default ellipse from a bivariate normal distribution, Figure 2.7, p. 38

Find `lessR` and `ggplot2` examples of adding an ellipse to a scatterplot from Chapter 2. Here specify multiple ellipses shown in Figure 5.2.

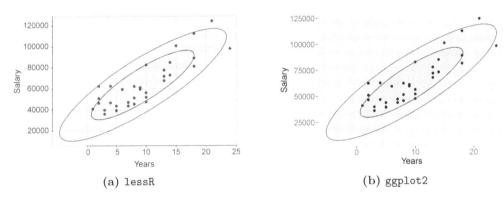

(a) `lessR`   (b) `ggplot2`

Figure 5.2: Multiple ellipses.

The instructions for creating the scatterplots in Figure 5.2 follows.

R Input *Two Ellipses:* Years *and* Salary - d *data frame*
data: d <- Read("http://lessRstats.com/data/employee.csv")

```
lessR: Plot(Years, Salary, ellipse=c(.68, .95))
ggplot2: ggplot(d, aes(Years, Salary)) + geom_point() +
         stat_ellipse(type="norm", level=.95,
           geom="polygon", color="gray20", size=.4, alpha=.05) +
         stat_ellipse(type="norm", level=.68,
           geom="polygon", color="gray20", size=.4, alpha=.05)
```

enhance parameter:
Create an enhanced
scatter plot as in
Figure 2.8, p. 40

For `lessR`, add the ellipse layer to the scatterplot from the `Plot()` function with the parameter `ellipse`, set to `TRUE` for a 0.95 confidence ellipse[1]. Or set `ellipse` to one or more specific values between 0 and 1, here with the base R `c()` function that defines a vector of values. The color of the ellipse defaults to a light translucent fill consistent with the current color theme. As illustrated in Chapter 2, `lessR` also provides the `enhance` option, which when set to `TRUE` adds an ellipse as well as other enhancements to the scatterplot.

c() function,
Section 1.2, p. 6

For `ggplot2`, explicitly add the ellipse with `stat_ellipse()`. The `level` parameter specifies the confidence level.

An alternative visualization specifies many ellipses and removes the points from the plot to focus on the underlying pattern suggested by the individual points. Here display the overlapping translucent interiors of the 0.05, 0.10, 0.15 ... 0.90, 0.95 data ellipses. Figure 5.3 results, which represent the idealized bivariate relationship estimated from the data, an informal, approximate display of the joint densities.

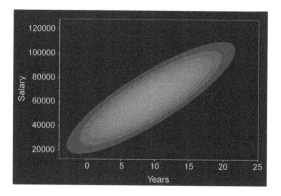

Figure 5.3: Cumulative translucent ellipses from `lessR` function `Plot()`.

The visualization in Figure 5.3 follows from two function calls. One function, `style()`, changes the style, applicable to subsequent visualizations. The other function plots the sequence of different ellipse interiors.

```
R Input Many Ellipses: Years and Salary - d data frame
data: d <- Read("http://lessRstats.com/data/employee.csv")

lessR: style("gray", sub_theme="black", ellipse_color="off", grid_lwd=0)
       Plot(Years, Salary, size=0, ellipse=seq(0.1,0.9,0.1))
```

Specify `size=0` to suppress the points. For additional impact, Figure 5.3 displays in grayscale with a black background, set by `style()`. Also with `style()`, turn off the color of the ellipse, that is, set to transparent. Turn off grid lines by setting their width to 0. Setting the `grid.color` to `"off"` accomplishes the same effect. To

---

[1]The `lessR` function `Plot()` obtains an ellipse from Murdoch and Chow's (2018) `ellipse` package with the function of the same name.

generate nine ellipses at nine different confidence levels, the base R `seq()` function generates a sequence of integers in this example from 0.1 to 0.9 in increments of 0.1.

seq() base R function: Generate a list of numbers by increments from a starting to an ending value.

To generate even more contrast between the ellipses and the background, change the style from `"gray"` to `"orange"` with `sub_theme` also set to `"black"`.

5.1.2 Line of Best Fit

In addition to the ellipse, another useful summary of a relationship displays a line of best fit, either non-linear or linear. For a non-linear fit, R provides `loess()`, the `lessR` and `ggplot2` default for the best-fit line. Figure 5.4 illustrates the loess smoothing of the scatterplot with the accompanying default 0.95 confidence region.

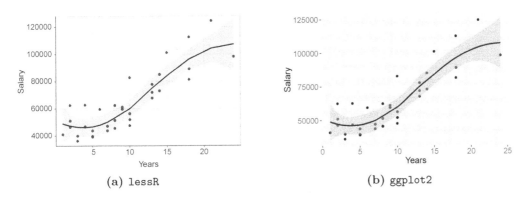

(a) `lessR`   (b) `ggplot2`

Figure 5.4: Loess best-fit lines with two-standard errors confidence intervals.

For `lessR`, obtain the line and confidence band with the `fit` parameter set to `TRUE`. For `ggplot2`, obtain the same with `geom_smooth()`.

R Input *Fit line and confidence intervals:* Years *and* Salary

data: d <- Read("http://lessRstats.com/data/employee.csv")

---

lessR: Plot(Years, Salary, fit=TRUE)

ggplot2: ggplot(d, aes(Years, Salary)) + geom_point() +
        geom_smooth(color="black")

Figure 5.5 illustrates the least-squares best-fit line with the corresponding confidence band. This visualization does not include the scatterplot plots that represent the data from which to calculate the fit-lines.

line types,
Section 10.4.1, p. 226

With `lessR`, suppress the points with `size=0`. For `ggplot2`, do not reference `geom_points()`. To display the linear model least-squares line, set `fit="lm"` as an argument to `Plot()` for `lessR`, and for `ggplot2`, set `method=lm` as an argument to `geom_smooth()`. To not display the confidence bands, for `lessR` set `fit_se` to `FALSE`, and for `ggplot2` set `se=FALSE` when calling `geom_smooth()`.

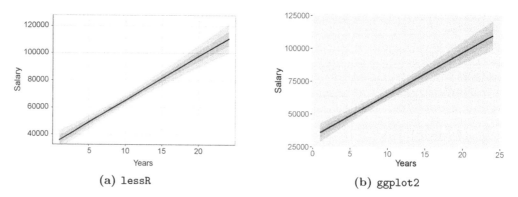

(a) <code>lessR</code>   (b) <code>ggplot2</code>

Figure 5.5: Least-squares best-fit lines with 0.68 and 0.95 standard error confidence intervals.

R Input *Least-squares Fit:* Years *and* Salary

data: d <- Read("http://lessRstats.com/data/employee.csv")

---

lessR: Plot(Years, Salary, size=0, fit="lm", fit_se=c(0.68, 0.95))

ggplot2: ggplot(d, aes(Years, Salary)) +
        geom_smooth(method=lm, level=.68, color="black") +
        geom_smooth(method=lm, level=.95, color="black")

The relevant `lessR` `style()` parameters are `fit.color` for the color of the fit line, `fit.lwd` for the width of the fit line, and `se.fill` for the color of the fill between the upper and lower confidence interval lines. Indicate one or more other confidence levels with the `fit_se` parameter.

For `ggplot2`, `color` specifies the color of the fit line, as illustrated, and `fill` specifies the fill color. Also, `linetype` specifies the type of line, such as `"dashed"` and `"dotted"`, and `size` specifies the width of the line. For `geom_smooth()`, specify confidence level with the `level` parameter.

Outlier Detection

Outlier identification: set `Plot()` parameter **enhance** to TRUE, Section 2.4, p. 38

Another task identifies and then labels potential outliers in the two-dimensional plot. `lessR` provides several parameters to accomplish this identification. The points so identified are both labeled and, for non-gray scale themes, displayed in red by default.

Mahalanobis distance: Measure of distance between two points across *n*-dimensions based on standard deviation and correlation.

The Quick Start section from Chapter 2 illustrates `lessR` outlier identification with the **enhance** parameter. `Plot()` relies upon the scale-invariant multivariate measure of distance called *Mahalanobis distance*. This distance measure assesses two criteria. First, for each variable, how many standard deviations does a point lie from the center of the multivariate distribution. Define the center of the scatterplot as the point with coordinates of the means of each variable. Second, how much do the variables correlate with each other?

The `Plot()` parameter `out.cut` labels the specified number of points that have the largest values of Mahalanobis distance, interpreted in the context of multivariate normality. Specify the parameter `out.cut` as a count or as a proportion. Or, invoke the parameter `out_MD`, which identifies all points with a Mahalanobis distance larger than the specified `out_MD` value. The default value of `out_MD` is 6 when `enhance` is set to `TRUE`, as shown in Figure 2.8.

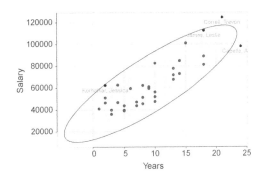

Figure 5.6: `Plot()` outlier analysis.

enhanced scatterplot,
Figure 2.8, p. 40

For Figure 5.6, `out.cut` is set to 0.10, so 10% of the 37 data values, rounded up to 4, with the highest values of Mahalanobis distance labeled.

R Input

```
data: d <- Read("http://lessRstats.com/data/employee.csv")
lessR: Plot(Years, Salary, ellipse=0.95, out.cut=.10)
```

To specify a variable other than row names for which to label the points, set the parameter `out_ID` to the variable name, without quotes.

correlation analysis,
Listing 2.3, p. 39

In addition to the statistical analysis of the correlation that `Plot()` provides, shown in Chapter 2, the outlier analysis also displays the values of Mahalanobis distance for the outliers, illustrated in Figure 5.1. Also shown are the next three largest values to aid in establishing a meaningful threshold that separates outliers from the remaining data values.

```
  MD               ID
 -----           -----
 8.14     Correll, Trevon
 7.84       Capelle, Adam
 5.63   Korhalkar, Jessica
 5.58       James, Leslie

 3.75         Hoang, Binh
 3.10       Skrotzki, Sara
 2.95       Billing, Susan
 ...               ...
```

Listing 5.1: `Plot()` Outlier analysis with Mahalanobis distance.

Of course, `ggplot2` could also generate the equivalent of Figure 5.6. To do so, however, would require the programming to calculate and sort the Mahalanobis distance of each point from the center, and then label the specified number of points according to their Mahalanobis distances.

5.1.3 Annotate

Often it is of interest to identify one or more specific points in the scatterplot.

Label all the points

One way to identify the individual points is with an identifier such as the data frame's row name. The result appears in Figure 5.7.

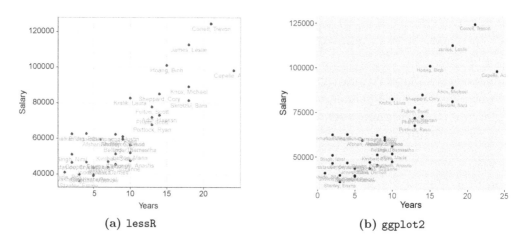

(a) `lessR`   (b) `ggplot2`

Figure 5.7: Scatterplot with each point identified by its row name.

Annotate a scatter plot for `lessR` with the `Plot()` add parameter, which also applies to `BarChart()` and `Histogram()`. Various values of **add** indicate different types of annotations, such as text or rectangles. The value of `"labels"` indicates to label all the points with the corresponding row names of the traditional base R data frame that contains the data for the plotted variables.

The `ggplot2 geom_text()` function adds text at the coordinates for values of the plotted variables. Use the base R `row.names()` to use the row names of the data frame as labels instead of a variable from the data frame, with the name of the data frame as its argument.

R Input *Scatterplot with Labels:* Years *and* Salary
data: d <- Read("http://lessRstats.com/data/employee.csv")

---

lessR: Plot(Years, Salary, add="labels")
ggplot2: ggplot(d, aes(Years, Salary)) + geom_point() +
        geom_text(aes(label=row.names(d)), size=3,
        color="darkgray", vjust=1.5)

For `lessR`, the relevant `style()` parameters for the labels are `add_color` and `add_cex` for color and size. The `size` and `color` parameters for `geom_text()` control their respective attributes. The `vjust` parameter adjusts the plotted text upward for a positive value so that the label does not over-plot the point itself. `ggplot2` also provides a `check_overlap` parameter to `geom_text()` that, when set to `TRUE`, does not display a label if the label would plot over another label.

Arbitrary annotations

Annotations can include arbitrarily placed geometric objects such as line segments and rectangles, as well as arbitrarily placed text. The scatterplot in Figure 5.8

includes text added at coordinate <21,105000> and a shaded rectangle that begins at <14,95000> and ends at <24.5,126000>.

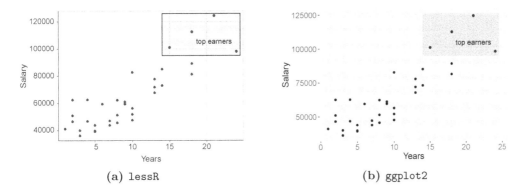

(a) lessR   (b) ggplot2

Figure 5.8: Text and rectangle annotations.

For lessR, optionally modify three of the add parameters with style(). Increase the size of the text message from the default of 0.6 to 1.0 with add_cex. Change the color of the rectangle border from the default of a medium gray to black with add_color. The default fill is a dark gray with no transparency. Here change the transparency to 0.9 to provide a high level of transparency.

The Plot() parameter add specifies the annotations, one annotation per value. Here specify two values. Positioning the text requires only a single coordinate, here specified with $x1$ and $y1$ to define the first values of the respective $x1$ and $y1$ vectors defined with the base R c() function. Define the two coordinates to specify the placement of the rectangle as the second values of the $x1$ and $y1$ vectors, and specify the sole values of $x2$ and $y2$.

Place each object with one to four corresponding coordinates, the required coordinates to plot that object, as shown in Table 5.1. x-coordinates may assume the value of "mean_x" and y-coordinates may assume the value of "mean_y".

Value	Object	Required Coordinates
"text"	text	x1, y1
"rect"	rectangle	x1, y1, x2, y2
"line"	line segment	x1, y1, x2, y2
"arrow"	arrow	x1, y1, x2, y2
"v.line"	vertical line	x1
"h.line"	horizontal line	y1
"means"	horiz, vert lines	

Table 5.1: lessR annotations and their corresponding coordinate fields.

The values of parameter add specifies the annotation objects. For a single object, enter a single value. Then specify the value of the needed corresponding coordinates, as defined in the above table. For multiple placements of that object, specify vectors of corresponding coordinates. To annotate multiple objects, specify multiple values for add as a vector. Then list the corresponding coordinates, for up to each of four coordinates, in the order of the objects listed in add.

For `ggplot2`, add text strings and geometric objects to the visualization with the `annotate` function. The previously illustrated `geom_text()` provides labels from values of a variable in the data table. Specify the type of annotation, here `"text"` and `"rect"`. Positioning the text requires a single x-y coordinate, with two coordinates needed to position the rectangle.

R Input *Scatterplot with Annotations:* Years *and* Salary

data: d <- Read("http://lessRstats.com/data/employee.csv")

---

lessR: style(add.cex=1, add.color="black", add.trans=.9)
    Plot(Years, Salary, add=c("top earners", "rect"),
      x1=c(21, 14), y1=c(105000, 95000), x2=24.5, y2=126000)
ggplot2: ggplot(d, aes(Years, Salary)) + geom_point() +
  annotate("text", x=21, y=105000, label="top earner", size=4.5) +
  annotate("rect", xmin=14, xmax=24.5, ymin=95000, ymax=126000, alpha=.2)

Specify each annotation in `ggplot2` as a distinct layer with separate function calls. For `lessR`, accomplish all the specifications with a single parameter, `add`. For multiple annotations, present the coordinates as vectors, with the required coordinates following the order of the values of the `add` parameter. Additional `lessR` annotation options include `"line"`, `"arrow"`, `"v.line"` for a vertical line, `"h.line"` for a horizontal line, and `"means"` for lines plotted at the means of the respective axes. Additional `ggplot2` annotations include `"segment"` specified by coordinates x, `xend`, y, and `yend`. The object `"pointrange"` is a line segment with a point in the middle. Specify a `"pointrange"` by coordinates x and y for the point, and either `xmin` and `xmax` or `ymin` and `ymax` for the line segment, depending on its orientation.

5.2   Consideration of a Third Variable

grouping variable: Categorical variable that defines groups.

How does the relationship of two continuous variables x and y vary across levels of a third, categorical variable? To visualize, plot one scatterplot for each level (group) of the third variable, the *grouping variable*. Plot the scatterplots either on the same panel, or on separate panels as a Trellis plot.

map data values, Section 10.2.1, p. 216

5.2.1   Map Data from a Grouping Variable to Aesthetics

map data values into a visual aesthetic: Color, size, or shape of a plotted object depends on the value of another variable.

To visualize multiple scatterplots with a single panel, *map values* of the grouping variable into a visual aesthetic such as color or size. Instead of setting the color or size of the plotted points, display the aesthetic according to the value of a third variable, the grouping variable.

Map to one visual aesthetic. Consider the relationship between *Years* employed at a company and annual *Salary* across the levels of *Gender*, shown in Figure 5.9. Map the value of *Gender*, with values of Male and Female in the data, into the color aesthetic, here constrained to grayscale. The relationships appear to be approximately similar, with a higher concentration of Females hired more recently.

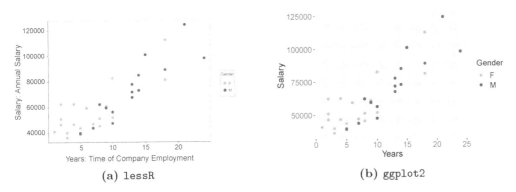

(a) `lessR`   (b) `ggplot2`

Figure 5.9: Scatterplot of two continuous variables at two levels of a categorical variable mapped into color.

The `lessR` `by` parameter specifies the grouping variable to plot values on the same panel. If grayscale is operational from a previous call to `style("gray")`, then by default `Plot()` implicitly taps into the `lessR` function `getColors()` to access the sequential scale `"grays"` to plot the levels. That is, with grayscale set, `Plot()` sets the `fill` parameter to `"grays"`.

sequential scale "grays", Section 10.2.2, p. 218

For `ggplot2`, the parameter `color` maps both edge color and fill of the points into the specified variable.

> **R Input** *Scatterplot with a grouping variable*
> *data*: `d <- Read("http://lessRstats.com/data/employee.csv")`
> ---
> *lessR*: `Plot(Years, Salary, by=Gender)`
> *ggplot2*: `ggplot(d, aes(Years, Salary)) + geom_point(aes(color=Gender)) +`
> `        scale_color_gray(start=0.3, end=0.7)`

qualitative palette, Section 10.2.1, p. 214

Default colors. Consider a nominal scale grouping variable, such as *Gender*. For `lessR`, without grayscale, the first two default colors are shades of blue and brown, from the default qualitative scale `"colors"`. The first two default colors for `ggplot2` are aqua and orange. For an ordinal grouping variable, `lessR` `getColors()` generates a sequential palette consistent with the color theme, such as `"blues"` for the default color theme of `"colors"`, and `"grays"` for the `"gray"` color theme. The default ordinal variable `ggplot2` palette is the viridis palette.

sequential palette, Section 10.2.2, p. 218

nominal variable, Section 1.2.3, p. 11

ordinal variable, Section 1.2.3, p. 10

Unlike `lessR`, `ggplot2`'s color theme `theme_gray` chooses colors from its qualitative palette. To display entirely in grayscale invoke the `scale_color_gray()` function. This example moderates the default range of luminance starting at 0.2 and 0.8 by restricting the range somewhat, instead of starting at 0.3 to 0.7.

viridis palette, Section 10.2.2, p. 220

HCL color space, Section 10.1.1, p. 212

Custom colors. A useful expression of color is with the HCL color space, according to the properties hue, chroma (saturation), and luminance (brightness). Particularly for nominal variables, usually choose custom colors of two different hues but the

equal intensity colors, Section 10.2.1, p. 215

same values of chroma and luminance so that all colors display at the same intensity. The `lessR` function `getColors()` provides a straightforward means to obtain colors of equal chroma and luminance.

fill and color parameters, Section 10.1.2, p. 213

In this example, set the first two colors of the qualitative palette to a lighter, more pastel version by increasing from their default values, chroma, `c`, to 80 and lightness, `l`, to 70 with `n` set to two colors. Specify `lessR` custom bar colors with the `fill` parameter, and custom bar border colors with the `color` parameter. For `ggplot2`, specify specific discrete colors with `scale_color_manual()` according to the `values` parameter.

> **R Input** *Scatterplot with grouping variable plotted with custom colors*
> *data*: d <- Read("http://lessRstats.com/data/employee.csv")
> *colors*: myClrs <- getColors(n=2, c=80, l=70)
> ───
> *lessR*: Plot(Years, Salary, by=Gender, fill=myClrs)
> *ggplot2*: ggplot(d, aes(Years, Salary)) +
>   geom_point(aes(color=Gender)) +
>   scale_color_manual(values=myClrs)

pre-defined color palettes with `getColors()`, Figure 10.3, p. 212

To generate custom sequential color palettes of a single hue for ordinal categorical variables, `getColors()` accesses 12 pre-defined sequential color palettes of 30-degree increments around the color wheel, with names such as `"reds"` and `"emeralds"`. The default hue for the palette follows from the current color theme, the default, or set with `style()`. Specify one of the 12 palettes from the color wheel. To change the default values of `c` and `l`, add these values to the `getColor()` function call.

Here generate two red colors with luminance varying from 70 to 35.

> *lessR*: myClrs <- getColors("reds", n=2, l=c(70,35))

Then plot with the `lessR` parameter `fill` or the `ggplot` function `scale_color_manual()`, set to the vector of generated colors, here *myClrs*.

Map to two visual aesthetics. Values of a third variable can map into different visual aesthetics for the same visualization. In Figure 5.10 example, map values of *Gender* into both the color and the shape of the plotted point. The plotted object is a letter, "F" or "M", for the respective genders, also plotted in different colors, here two shades of gray.

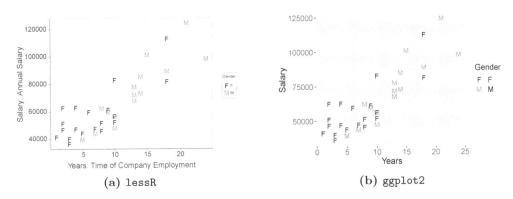

(a) lessR   (b) ggplot2

Figure 5.10: Scatterplot of two continuous variables at two levels of a categorical variable mapped into shape and color.

The relevant parameter for plotting letters for both `lessR` and `ggplot2` is `color`. The `color` parameter is not needed except to override the default grayscale values if `style("grays")` had been previously set. For `lessR`, also specify the `shape` parameter. For `ggplot2`, set the `shape` parameter equal to *Gender* within the `aes()` function, and then define the specific shapes with a function call to `scale_shape_manual()`. Use the `values` parameter to assign the specific shape to each value of *Gender*.

R Input

```
data: d <- Read("http://lessRstats.com/data/employee.csv")

lessR: Plot(Years, Salary, by=Gender,
           color=c("gray5", "gray55"), shape=c("F","M"))
ggplot2: ggplot(d, aes(Years, Salary)) +
           geom_point(aes(color=Gender, shape=Gender)) +
           scale_color_manual(values=c(F="gray5", M="gray55")) +
           scale_shape_manual(values=c(F="F", M="M"))
```

5.2.2   Trellis (Facet) Scatterplots

Applied to a scatterplot, *Trellis Graphics* plots the scatterplot for each level of a grouping variable in a separate plot called a *panel*, as in Figure 5.11.

With `lessR`, specify a separate panel for each level of the grouping variable with the `by1` parameter. With `ggplot2`, invoke the `facet_grid` function to create what `ggplot2` refers to as facets, that is, panels of a Trellis plot. As seen from Figure 5.11, the `lessR` standard error bands do not currently transfer over to the Trellis graphics panels.

Trellis graphics: A rectangular grid of panels of the same variables for each level of one or more categorical variables.

Trellis graphics, Figure 2.12, p. 44

R Input *Trellis (facet) scatterplots*

```
data: d <- Read("http://lessRstats.com/data/employee.csv")

lessR: Plot(Years, Salary, by1=Gender, fit="lm")
```

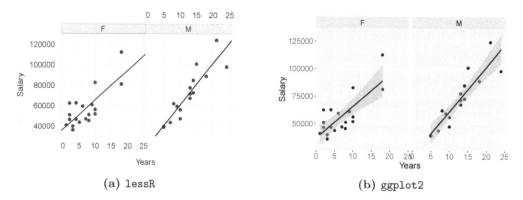

(a) `lessR`   (b) `ggplot2`

Figure 5.11: Trellis (facet) scatterplot of two continuous variables.

> *ggplot2*: `ggplot(d, aes(Years, Salary)) + geom_point() +`
>   `geom_smooth(method="lm", color="black") +`
>   `facet_grid(cols=vars(Gender))`

For `lessR`, control the layout of the panels by choosing one of the `n_col` and `n_row` parameters. For example, to stack the two panels into a single column set `n_col` to 1. For `ggplot2`, use the parameter `cols` to indicate to place the panels in columns. Refer to the variable of interest, here *Gender*, as an argument to the `vars` function. The parameter `rows` indicates to stack the panels on top of each other.

Trellis plots can also be constructed with two grouping variables. For `lessR`, invoke parameter `by1`, and then `by2` for the second grouping variable. For the `ggplot2` `facet_grid` function, specify both a `cols` parameter and a `rows` parameter.

Another possibility, illustrated in Figure 5.12, specifies one grouping variable to display on the same panel, and a second grouping variable to specify a Trellis plot. The variable that displays within each panel, *JobSat*, has three different levels. The grouping variable that defines the separate panels is *Gender*.

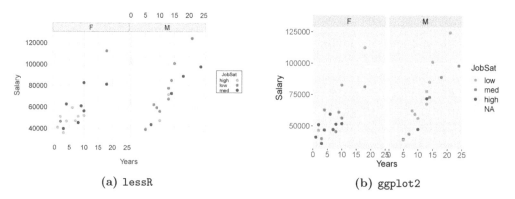

(a) `lessR`   (b) `ggplot2`

Figure 5.12: Scatterplot of two continuous variables with a categorical conditioned variable and a categorical grouping variable.

For `lessR`, use both the `by` and `by1` parameters to generate within-panel and between-panel scatterplots. For `ggplot2`, repeat the steps for generating a single-panel and multi-panel scatterplots across levels of a grouping variable: Add the `color` parameter to the `aes()` function for the single-panel scatterplots, and then add the `facet_grid()` function for the multi-panel scatterplots.

R Input *Multi- and single-panel scatterplots combined*

data: d <- Read("http://lessRstats.com/data/employee.csv")

lessR: Plot(Years, Salary, by1=Gender, by=JobSat)

ggplot2: ggplot(d, aes(Years, Salary)) +
        geom_point(aes(color=JobSat)) +
        facet_grid(cols=vars(Gender)) +
        scale_color_gray(start=0.6, end=0.3, na.value-"white")

This process easily generalizes to all combinations of panels for Trellis plots for two other grouping variables.

5.2.3 Map a Third Continuous Variable into a Visual Aesthetic

The previous examples in this section mapped a categorical variable into a visual aesthetic. The same concept applies to a continuous variable. Figure 5.13 illustrates the mapping of the test scores of the variable *Pre*, which range from 59 to 100, into the size of the plotted points. To indicate the scale of the plotted sizes, `ggplot2` provides a legend of various point sizes. In contrast, `lessR` indicates the sizes of the smallest point and the largest point in the title, also labeling these points on the scatterplot.

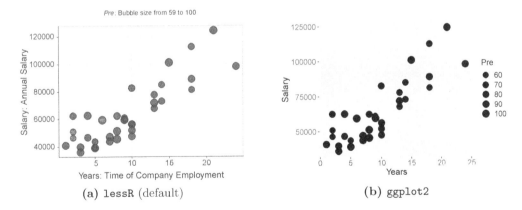

(a) `lessR` (default)   (b) `ggplot2`

Figure 5.13: Scatterplot of two continuous variables mapping a third continuous variable into the size of the plotted points represented by their area.

Mapping a continuous variable to an aesthetic share the same `ggplot2` mechanics as mapping a categorical variable. Within `aes()` set the visual aesthetic, such as `size`, to the variable that maps into the aesthetic. For `lessR`, do the same, set the parameter `size` to the variable to be mapped. The `lessR` parameters `by`, `by1`, and `by2` from the previous examples apply only to categorical variables.

The bubble plots in Figure 5.13 set the area of the bubble to the size, subject to a scaling factor so that all the bubbles comfortably display on the same panel. Accordingly, the value of the mapping variable of 0 results in a bubble of size 0. We perceive area better than the radius of the circle, so area is typically considered the better scaling for the mapping variable. That scaling is the lessR default. To obtain the scaling with ggplot2, add the call to the function scale_size_area().

R Input *Bubble plot with area of each bubble equal to scaled size*
data: d <- Read("http://lessRstats.com/data/employee.csv")

lessR: Plot(Years, Salary, size=Pre)
ggplot2: ggplot(d, aes(Years, Salary, size=Pre)) + geom_point() +
            scale_size_area()

The values of *Pre* only vary from 59 to 100, with the result that bubbles scaled as area with a fixed 0 are not much differentiated. Figure 5.14 presents an alternative set of bubble plots that exhibits more variation in the bubble sizes.

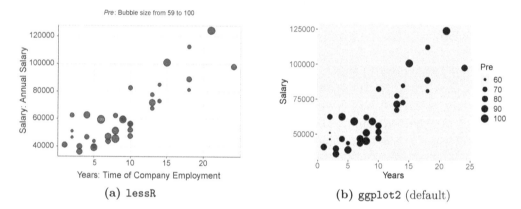

(a) lessR   (b) ggplot2 (default)

Figure 5.14: Scatterplot of two continuous variables mapping a third continuous variable into the size of the plotted points.

For lessR, the parameter **power** sets the relative size of the scaling of the bubbles to each other. The default value of 0.5 scales the bubbles, so that the area of each bubble is the scaled value of the corresponding sizing variable. A value of 0 leaves all bubbles the same size. A value of 1 scales the radius of the bubble to the value of the sizing variable, which increases the discrepancy of size between the variables, to more effectively accentuate differences. There is no practical upper bound for **power**, so its value can be varied continuously to explore different relative bubble sizes. For Figure 5.14, **power** equals 2.

To obtain a corresponding accentuation of the differences in bubble sizes with ggplot2, rely on its default, so do not specify the scale_size_area() function. There does not appear to be a parameter to manipulate that changes the relative bubble sizes.

R Input *Bubble plot with areas more differentiated*
data: d <- Read("http://lessRstats.com/data/employee.csv")

```
lessR: Plot(Years, Salary, size=Pre, power=2)
ggplot2: ggplot(d, aes(Years, Salary, size=Pre)) + geom_point()
```

For `lessR`, the parameter `radius` scales the bubbles to the radius of the largest displayed bubble in inches. The `lessR` default value for mapping a continuous variable to `size` is 0.10. To change the absolute bubble sizes in `ggplot2`, if scaling to area with a floor of 0, for the `scale_size_area()` function change the size of the parameter `max_size` from its default value of 6. Otherwise, for the function `scale_size_area()` change minimum and maximum values of the parameter `range` from its default of `c(1,6)`.

5.2.4   Plot Multiple Variables on the Same Panel

scatterplots based on groups, Figure 5.9, p. 113

A previous example displayed multiple scatterplots on the same panel by subsetting the data for the two continuous variables into groups and plotting a separate scatterplot for each group. In this section, plot multiple scatterplots on the same panel, but each scatterplot is of separate variables. Figure 5.15 illustrates with the variables *Pre* and *Post* on the *x*-axis and the variable *Salary* on the *y*-axis, also plotted with the two corresponding regression lines and standard error bands.

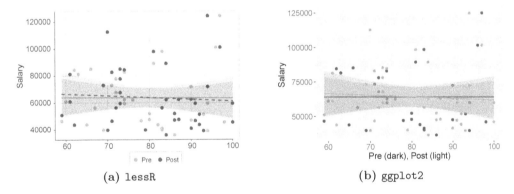

(a) `lessR`   (b) `ggplot2`

Figure 5.15: Scatterplot of two x-axis variables plotted against a y-variable.

To plot multiple variables on either the *x*- or *y*-axis, with `lessR` replace a single variable with a vector of multiple variables. To specify a regression line for each of multiple variables, invoke the parameter `fit`. For `ggplot2`, add each scatterplot with a `geom_point()` layer and each regression line with a `geom_smooth()` layer.

fit line, Section 5.1.2, p. 107

R Input *Scatterplot with two x-axis variables*

```
data: d <- Read("http://lessRstats.com/data/employee.csv")

lessR: Plot(c(Pre, Post), Salary, fit="lm")
ggplot2: ggplot(d) +
         geom_point(aes(Pre, Salary), color="gray30") +
         geom_smooth(aes(Pre, Salary), method="lm", color="gray30") +
         geom_point(aes(Post, Salary), color="gray60") +
```

```
        geom_smooth(aes(Post, Salary), method="lm", color="gray60") +
        labs(x="Pre (dark), Post (light)")
```

The *x*-variable in the call to `lessR Plot()` is c(Pre, Post), a vector of two variables, *Pre* and *Post*. The base R `c()` function provides the most general way to create a vector, to indicate multiple values. Plot the scatterplot and regression line separately for each indicated variable.

5.3   Inter-Relations of a Set of Variables

This section provides two techniques for viewing the relation of each variable to each other variable in the analysis: The scatterplot matrix and the heat map.

5.3.1   Scatterplot Matrix

scatterplot matrix (SPLOM): Matrix of pairwise scatterplots for a set of variables.

The scatterplot matrix (SPLOM) displays all the pairwise scatterplots of a set of variables, which may include additional information such as a fit lines. Figure 5.16 presents the `lessR` SPLOM with an added linear fit line for each scatterplot. A glance at the SPLOM indicates that *Salary* and *Years* have a strong correlation, whereas *Pre* is not related to either of these two variables.

Early versions of `ggplot2` contained a SPLOM function, but no more, though additional alternatives do exist. Figure 5.17 presents two different SPLOM's of the same data as in Figure 5.16 using functions from the `psych` package (Revelle, 2018), and the `GGally` package (Schloerke et al., 2018), which is based on `ggplot2`.

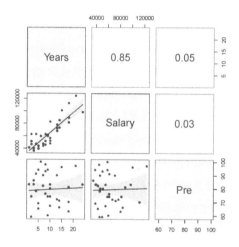

Figure 5.16: `lessR` scatterplot matrix with linear fit lines and standard errors.

With `lessR`, to obtain a scatterplot matrix use `Plot()`. For the x-parameter, the first argument passed to the function, specify a vector of numeric variables. The usual convention applies the default data frame name of *d*, or specify the data frame with the `data` parameter. The parameter `fit` for the best fitting summary line also applies, with the options of `"loess"` for non-linear fit and `"lm"` for linear fit. Many of the standard `lessR` visualization parameters also transfer over to the scatterplot matrix, for example, red plotted points for the color theme `"darkred"`.

fit parameter, `lessR`, Section 5.1.2, p. 107

pairs.panels() function, psych: generate a scatterplot matrix.

The `psych` package offers the function `pairs.panels`, which accepts the data as a data frame. To select a subset of variables from a data frame, choose from among the base R function `subset` with the parameter `select`, or the tidyverse function `subset`, or the base R subsetting convention with square brackets to specify specific

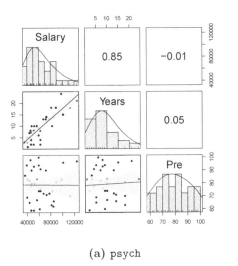

(a) psych

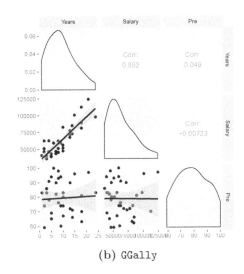

(b) GGally

Figure 5.17: Scatterplot matrices.

rows and/or columns of a data frame, in this example, *d*. Specify the rows before the comma, and the columns after. To select all the rows, do not specify any argument before the comma. To select the columns, specify a vector of variables after the comma. So,

subset a data frame,
Section 1.2.4, p. 13

```
d[,c("Salary", "Years", "Pre")]
```

defines a new data frame with all rows of data from the *d* data frame, but only retains the specified variables.

The remaining parameters of `pairs.panels` are straightforward. Setting `lm`, `ci`, and `density` to `TRUE` as in this example activates, respectively, the best-fitting least-squares line, the confidence interval band about each line, and densities down the diagonal. The `ellipses` option is turned off. The `hist.col` and `col` options specify, respectively, the color of the histogram bars and the fit line. The `cex.cor` parameter of 0.7 reduces the size of the displayed correlations from the default value of 1.

For `GGally`, the `ggpairs` function provides the SPLOM. Here, in the style of `ggplot2`, the first listed parameter is the data frame, with the variables selected as a vector specified by the `columns` parameter. The `lower` parameter with the list argument of `continuous`, set to `"smooth"`, indicates to include the linear fit line in each scatterplot.

ggpairs() function,
GGally: generate a
scatterplot matrix.

R Input *Scatterplot matrices*
data: d <- Read("http://lessRstats.com/data/employee.csv")

lessR: Plot(c(Years, Salary, Pre), fit="lm")
psych: pairs.panels(d[,c("Salary", "Years", "Pre")],
        lm=TRUE, ci=TRUE, density=TRUE, ellipses=FALSE,
        hist.col="gray80", col="black", cex.cor=.07)
GGally: ggpairs(d, columns=c("Years", "Salary", "Pre"),
        lower=list(continuous="smooth"))

The SPLOM reveals the relationships among a set of variables. Regression analysis provides a formal analysis of these relationships. The SPLOM as part of a comprehensive regression analysis is provided by the `lessR` function `Regression()`, abbreviated `reg()`.

> *comprehensive least-squares regression analysis including SPLOM*
>
> *lessR*: `reg(Salary ~ Years + Pre)`

For only a single predictor variable, `Regression()` displays the scatterplot of response variable and predictor variable, with the least-squares regression line plus the confidence interval and prediction interval bands about the line. For multiple regression, the same SPLOM as in Figure 5.16 is generated. These visualizations help explain the statistical results and contribute to variable selection. For example, Figure 5.16 reveals that in the corresponding multiple regression model predicting *Salary*, the predictor variable *Years* would be retained due to its strong relationship to *Salary* and the predictor variable *Pre* dropped from the model.

5.3.2   Heat Map of a Correlation Matrix

tile: Small plotted square colored as a mapped data value.

heat map: Replace each number in a matrix with a scaled tile.

Mach IV scale, Section 1.2.4, p. 17

correlogram: Visual display of a correlation matrix such as a heat map.

Correlation() function, `lessR`: Compute a correlation matrix or a single correlation.

A correlation matrix organizes all the pair-wise correlations among a set of variables into a square, symmetric matrix, a summary of the linear relationships among the variables. To search for patterns among this table of numbers, represent the correlations visually, here with a *heat map*, which replaces each correlation in a correlation matrix with a colored *tile*, a small square of color. The larger the correlation, the more intense the tile color. A correlation matrix visualized as a heat map is an example of a *correlogram* (Friendly, 2002).

The data in this example consist of 351 responses on a 6-pt Likert scale applied to the items on the 20 item Mach IV scale presented in the first chapter, gathered in a study of Machiavellianism beliefs (Hunter et al., 1982). The purpose of the study is common to many analyses of the items on a multi-item scale such as the Mach IV: Investigate the underlying patterns of the correlation coefficients of the items to potentially uncover sub-scales that measure related but yet distinct concepts.

Correlation matrix. The `lessR` function `Correlation()`, abbreviated `cr()`, computes the correlation matrix. Pass an entire data frame to the first parameter, x, to obtain the corresponding correlation matrix. If the data frame contains non-numeric variables, unlike the base R `cor()` on which `cr()` is based, `cr()` does not exit with an error, but instead deletes all the non-numeric variables with notification. The output of `cr()` is an R data structure called a `list` that consists of multiple components, or pieces of information. The output provides information such as missing data, as well as the correlation matrix per se, which is a component named R, a traditional name for a correlation matrix.

By default `cr()` computes the traditional Pearson correlations. Because this function relies on base R `cor()`, also can set the `cor()` parameter `method` to `"kendall"` for Kendall's *tau* correlation statistic or to `spearman` for Spearman's *rho* correlation statistic. The default approach to missing data is `"pairwise"` deletion, which uses all available data to calculate each correlation coefficient. To invoke listwise deletion, set the parameter `miss` to `"listwise"`, which means that if any data are missing in a row, the entire row of data is deleted. Finally, as with the base R `cor()` function, if a pair of numeric variables is passed to `cr()` as parameters x and y, `cr()` computes the single correlation coefficient. Unlike the base R function, `cr()` also provides the corresponding inferential analysis.

In this example, `Read()` reads the responses to the Mach IV items into the data frame d. For purposes of this illustration, to keep the output more compact, in the second step delete the last four items on the Mach IV scale with the base R function `subset()`. None of these items appeared on any of the sub-scales from (Hunter et al., 1982).

The correlation matrix of the remaining 16 variables, available as the list component R, identified by R, is computed with `cr()`, here saved into the matrix object also named R. Reference the data frame d in the call to `cr()` for completeness, though not necessary as it is the default data frame.

Get the data and compute the correlation matrix

```
lessR: d <- Read("http://lessRstats.com/data/Mach4.csv")
       d <- subset(d, select=c(m01:m16))
       R <- cr(d)$R
```

```
1  > R
2       m01 m02 m03 m04 m05 m06 m07 m08 m09 m10 m11 m12 m13 m14 m15 m16
3  m01 100  07  16 -08  17 -11 -13  22 -14 -15 -09  25  17 -03  11 -08
4  m02  07 100  00 -15  12 -04 -09  14 -18 -22  01  18  13 -02  25 -03
5  m03  16  00 100  07  06  24  24  06  16  15  10  04 -06  10 -01  10
6  m04 -08 -15  07 100 -09  09  14 -11  14  11  14 -18 -14  23 -01  13
7  m05  17  12  06 -09 100 -11 -10  10 -09 -02 -13  23  22  06  09 -19
8  m06 -11 -04  24  09 -11 100  52 -05  25  41  29 -14 -16  11 -15  24
9  m07 -13 -09  24  14 -10  52 100 -09  32  40  22 -16 -11  15 -17  21
10 m08  22  14  06 -11  10 -05 -09 100 -10 -07 -10  10  09 -12  06 -06
11 m09 -14 -18  16  14 -09  25  32 -10 100  25  16 -14 -13  06 -17  17
12 m10 -15 -22  15  11 -02  41  40 -07  25 100  15 -18 -08  15 -18  11
13 m11 -09  01  10  14 -13  29  22 -10  16  15 100 -20 -11  26 -02  28
14 m12  25  18  04 -18  23 -14 -16  10 -14 -18 -20 100  22 -11  08 -15
15 m13  17  13 -06 -14  22 -16 -11  09 -13 -08 -11  22 100  12  16 -03
16 m14 -03 -02  10  23  06  11  15 -12  06  15  26 -11  12 100  01  22
17 m15  11  25 -01 -01  09 -15 -17  06 -17 -18 -02  08  16  01 100 -01
18 m16 -08 -03  10  13 -19  24  21 -06  17  11  28 -15 -03  22 -01 100
```

Listing 5.2: Correlation matrix of the 16 Mach IV items (to save space `lessR` rounds the correlations to two digits with the decimal points removed).

Listing 5.2 presents R, the table of correlations for which to search for the underlying structure. Relatively sophisticated statistical analyses of a correlation matrix such as exploratory and confirmatory factor analysis are typically applied to uncover the latent structure of a multi-item scale. Data visualization, here in the form of a heat map, serves as a useful, complementary tool to the statistical analyses. The heat map facilitates the potential emergence of visual patterns that indicate a sub-structure, such as the full multi-item scale.

seriation: A linear ordering of variables to indicate underlying structure.

Seriation. To enhance the visual recognition of patterns with the heat map, first, order the correlations by their similarity. One strategy reorders the variables into a linear order to better reveal the underlying structure in the matrix, a process called *seriation* (Liiv, 2010). Without reordering, the patterns are likely obscured, so seriation is usually a necessary step in the display of a correlation matrix heat map. There are many seriation algorithms, of which two follow.

hclust() function, **base** R: Hierarchical cluster analysis.

1. Hierarchical cluster analysis groups the variables into clusters, which results in a linear ordering of the variables, the items in this example. The analysis proceeds by initially designating each variable as its own cluster. Then combine the closest two clusters into a single cluster, repeating the process until all the variables form a single cluster. Although many ways exist to define what is meant by the closest cluster, the de-

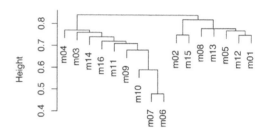

Figure 5.18: Hierarchical cluster dendrogram for 16 Mach IV items.

dendrogram: Visualization of relationships as a binary tree, from first split at the top, continuing down to the leaves, the variables.

fault method by the base R hierarchical clustering function **hclust()** generally works well on correlation matrices.

The hierarchical cluster analysis organizes the groups of variables into a binary tree called a *dendrogram*, illustrated in Figure 5.18 computed with the default method from **hclust()**. The top of the binary tree divides all variables into the two most dissimilar groups. Each split is called a node. Then the second level of the tree splits all the elements of each of the two different groups, into the two most dissimilar groups, leaving four nodes and clusters. These binary splits continue until only the original elements remain, the leaves (items) of the tree, with no more splitting possible.

The heat map in Figure 5.19a displays the results of the hierarchical cluster analysis. The heat map functions do not require the dendrogram. The following code, however, shows the simplicity of obtaining the dendrogram from the correlation matrix R.

Dendrogram of a correlation matrix

base R: `plot(hclust(as.dist(1-R)))`

The `hclust()` function analyzes a dissimilarity matrix, whereas the correlation matrix is a similarity matrix. To convert the correlations to dissimilarities, subtract all correlation coefficients from 1. `hclust()` also needs the matrix in the format created by the distance function `dist()`. The function `as.dist()` converts to that format. Then analyze with `hclust()` and plot the dendrogram.

fast optimal leaf ordering: Seriation procedure that maximizes the sum of the similarity of adjacent variables.

2. Bar-Joseph, Gifford, and Jaakkola (2001) present a seriation algorithm, *fast optimal leaf ordering*, that further refines the output from the hierarchical cluster analysis. Given this binary tree, the optimal leaf ordering is the order that maximizes the sum of the similarity of adjacent variables, while retaining the original structure of the binary tree. Nodes and corresponding sub-trees can be flipped to list the

elements in a different order, but the structure of the binary tree remains unchanged with the same hierarchy, just re-ordered. Place the most similar variables adjacent to each other to highlight any underlying patterns that may emerge. Highly correlated variables appear in the middle of a cluster, and more marginal variables appear at the borders. The heat map in Figure 5.19b displays the results of optimal leaf ordering seriation.

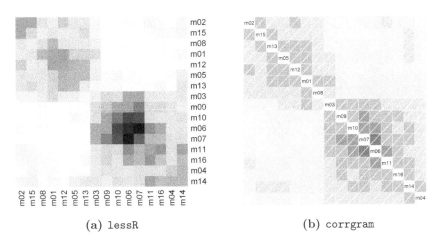

          (a) `lessR`   (b) `corrgram`

Figure 5.19: Heat maps of a correlation matrix from optimal leaf ordering seriation.

Obtain the ordered heat maps with the `lessR` function `corReorder()`, and with the function `corrgram()` from the `corrgram` package (Wright, 2018). Again, for `lessR`, grayscale has been set for all output with `style("gray")`.

> **R Input** *Ordered heat map*
> *data*: *data to correlation matrix steps shown above*
> ---
> *lessR*: corReorder(R, order="hclust")
> *corrgram*: corrgram(R, order="OLO",
>   col.regions=colorRampPalette(c("white", "gray85", "black")))

corReorder(),
function, `lessR`:
Reorder a correlation
matrix.

corrgram(),
function, `corrgram`:
Reorder a correlation
matrix.

<u>`lessR`</u> heat map. `corReorder()` generates a heat map of the reordered matrix consistent with the current color theme, relying upon the base R function `heatmap()`. For example, for the default `lessR` color theme of `"colors"`, to maximize the visual appearance of the tiles across the range of values in the correlation matrix, `corReorder()` displays the heat map with a divergent HCL color scale with brown for the smaller values and blue for the larger values. All the color themes result in divergent scales except when grayscale is set, as in Figure 5.19a, where `corReorder()` defaults to a sequential palette across the full range of grayscale.

The diagonal elements of the correlation matrix differ from the off-diagonal elements because the diagonal correlations all equal 1.0, that is, each variable perfectly correlates with itself. To have the diagonal of the heat map blend in better with the remaining tiles so as to improve pattern recognition, analogous to the computation of a communality in a factor analysis, `corReorder()` sets the diagonal element for each variable for the heat map as the average of the two adjacent variable correlations off of the diagonal. For the first and last variable in the matrix, the value of the one adjacent variable replaces the 1.0. To display the original diagonal values, set `diagonal.values` to `FALSE`.

`corReorder()` also provides for a manual reordering of the correlation matrix, including variable deletion, which allows manual fine-tuning of the initial results from an algorithm, or experimentation to view the results of a specific recording. Indicate the specified ordering by providing a vector of ordered variable names to the parameter `vars`, which also sets the `order` parameter to `"manual"`.

For example, the heat maps in Figure 5.19 reveal the strongest correlation, between *m07* and *m06*, so perhaps place those two items and two other highly related items, *m09* and *m10*, at the beginning of the heat map. Items *m03* and *m08* appear relatively isolated. The following code accomplishes the re-ordering. Leaving *m03* and *m08* out of the list deletes them from the heat map.

Manual reordering of the correlation matrix

```
lessR: corReorder(R, vars=c(m09,m10,m06,m07,m11,m16,m04,m14,m02,m15,
                            m01,m12,m05,m13))
```

`corReorder()` also implements Hunter's (1973) seriation procedure when `order` is set to `"chain"`. The procedure is simple but surprisingly effective for correlation and related matrices. The first variable is that with the largest sum of squared correlations across all the variables, the variable that can be considered to be the most related to all the variables in the set. Then the variable that has the highest correlation with the first variable is listed second, and so forth. When applied to the Mach IV data, this simple procedure identified the sub-groups about as well as the more formal implementations shown in Figure 5.19.

The output of `corReorder()` with default values is the re-ordered correlation matrix from the hierarchical cluster analysis, here not displayed at the R console, but saved to a specified output object for later processing.

Save the reordered correlation matrix

```
lessR: R <- corReorder(R)
```

In this example, save the reordered correlation matrix over the input matrix. To retain both matrices save to a name with a different name than the input correlation matrix.

corrgram heat map. The default colors of the `corrgram()` function are salmon to blue, but the grayscale added with the `col.regions` parameter works about as well. Implement the optimal leaf ordering from the **seriation** package (Hahsler, Hornik, & Buchta, 2008), which also provides many other seriation methods, some of which **corrgram** also provides with its `order` parameter. Set `order` to `"HC"` to stop at the hierarchical clustering without further seriation. A seriation algorithm with a similar goal as optimal leaf ordering but with a less formal basis (Gruvaeus & Wainer, 1972) is also available with `order` set to `"GW"`. Finally, obtain the ordering based on principle components with the value of `"PCA"`. There is no default ordering, so if `order` is not specified the original order is retained.

The parameter `col.regions` specifies the colors for generating the heat map. `corrgram()` uses the base R function `colorRampPalette()` to generate its color palettes.

corrplot heat map. The `corrplot` package provides a variety of correlograms with its `corrplot()` function. With the `method` parameter set to `"number"`, display each correlation literally as a shaded color, `"pie"` to display a mini pie chart, `"ellipse"` to display a confidence ellipse, and `"square"` for a shaded square for each correlation. Obtain the basic heat map with `"color"`. Obtain the same with the added shade line for negative correlations with `"shade"`, as in Figure 5.19b, which has the shade line for all correlations.

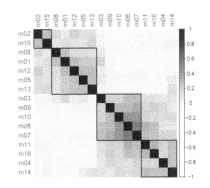

Figure 5.20: `Corrplot()` heat map correlogram with sub-scales.

As can be seen from Figure 5.20, `corrplot` also provides a unique feature, a rectangle around the correlations of groups of items based on the hierarchical cluster analysis given a specified number of clusters that the analyst determines from viewing preceding correlograms. After obtaining the first heat map, specify the number of rectangles to add to the plot to emphasize emerged patterns of correlations.

To generate Figure 5.20 beyond the input correlation matrix, use `method` to specify the heat map, `order` to specify hierarchical cluster analysis, and `addrect` to indicate to draw rectangles at the four cluster solution.

R Input *Ordered heat map*
data: `data to correlation matrix steps shown above`

```
corrplot: corrplot(R, method="color", order="hclust", addrect=4,
           tl.col="gray30",
           col.regions=colorRampPalette(c("white", "gray85", "black")))
```

The `tl.col` parameter indicates to display the column labels in gray, and the `col.regions` parameter specifies the defined grayscale for the generated heat map.

Revealed structure. The heat map delineates the underlying structure of the items of the Mach IV scale. Even with the grayscale heat maps, a clear pattern emerges. The items *m07, m06, m10,* and *m09* form the most defined sub-scale. From there, the next two items, *m02* and *m15,* cluster together, as do *m13, m05, m12, m01,* and to a lesser extent *m08.* At the other corner of the heat map, items *m14, m11, m16,* and to a lesser extent *m04,* cluster.

The appearance of four sub-scales from Figure 5.19 verifies the sub-scale membership to the four clusters with the `corReorder()` parameter `n_clusters`.

> *Generate clusters from the dendrogram*
>
> *lessR*: `corReorder(R, n_clusters=4)`

After viewing the ordered heat map and assessing the potential number of sub-scale, use `corReorder()` to implicitly invoke `cutree()` to obtain the hierarchical cluster analysis solution according to the specified four clusters. Find the output in Listing 5.3.

```
m01 m05 m08 m12 m13 m02 m15 m03 m06 m07 m09 m10 m04 m11 m14 m16
  1   1   1   1   1   2   2   3   3   3   3   3   4   4   4   4
```

Listing 5.3: Default hierarchical four-cluster solution.

When compared to the solution obtained from exploratory and confirmatory analysis (Hunter et al., 1982), with a formal test of fit, the heat map revealed a sub-structure remarkably close to the formal confirmatory factor analysis solution, which identified four components of Machiavellianism as assessed by Mach IV, summarized in Listing 5.4 that names the factors and lists a representative item for each.

```
Deceit: m06, Honesty is the best policy in all cases.
Flattery: m15, It is wise to flatter important people.
Immorality: m04, Most people are basically good and kind.[reversed scored]
Cynicism: m12, Anyone who completely trusts anyone else is asking for trouble.
```

Listing 5.4: Four Mach IV Machiavellian subdomains and sample items.

The heat map of the reordered correlation matrix performed well in recovering the underlying structure of the Mach IV scale. The only difference between the solutions of the cluster analysis and the formal confirmatory factor analysis is that the Items *m08* and *m14* did not appear in the final solution. The full factor analysis was performed on the full 20x20 Mach IV matrix. Repeating the cluster analysis on the full 20-item matrix, adding more clusters than only four to provide a place for the items that did not contribute to the confirmatory factor analysis solution, lead again to an excellent recovery.

The criterion of fit for the correlations of items on sub-scales in a confirmatory factor analysis, such that each item contributes to the measurement of only a single scale (factor), is not the size of the correlations, but their patterning (Gerbing & Anderson, 1988). The correlations for two items on the same sub-scale have proportional correlations across all other variables. The structure of the correlation matrix, more generally referred to as a covariance structure, is the basis for the statistical technique of confirmatory factor analysis. Directly evaluate the proportionality criterion for the items on sub-scales by calculating their coefficient of proportionality with the `lessR` function `corProp()`.

corProp() function, `lessR`: Transform a correlation matrix to a matrix of proportionality coefficients.

geom_tile() function, **ggplot2**: Construct a heat map composed of tiles.

ggplot2 heat map. For completeness, the `ggplot2` heat map is also presented. `ggplot2` does not directly provide for a reordering of the correlation matrix, so the `corReorder` function call in the following code saved the reordered matrix, replacing the original unordered *R* with the ordered version, then analyzed by `ggplot2`.

R Input *Ordered correlation matrix heat map*
data: `data to correlation matrix steps shown above`

---

lessR: `R <- corReorder(R)`

reshape2: `Rmelt <- melt(R, value.name="cor")`

ggplot2: `ggplot(data=Rmelt, aes(Var2, Var1, fill=cor)) + geom_tile() +`
`            scale_fill_gradient2(low="black", mid="white", high="black",`
`                midpoint=0, limit=c(-1,1)) + coord_fixed() +`
`            labs(x="", y="") + theme(axis.text.x=element_text(angle=90))`

Before the `ggplot2` function to create the tiled heat map, `geom_tile()`, can be invoked, the correlation matrix needs to be restructured. All the analysis functions in the tidyverse, including `ggplot2`, require the data to be in a specific form, observations in the rows and variables in the columns, called tidy data (Wickham, 2014). Convert the matrix from its present wide form, in which all the correlations for a variable appear in the same row, to the corresponding long form in which each correlation appears in its own row. This conversion can be accomplished with `stretch()` from the `corrr` package (Jackson, Cimentada, & Ruiz, 2019) if the correlation matrix is computed with its own `correlate()` function, or converted to the `corrr` data type. Here, show the conversion step-by-step to illustrate the needed data manipulations.

Tidyverse functions that convert data from wide to long form are `gather()` and, more recently, `pivot_long()` from the `tidyr` package (Wickham & Henry, 2019). Although these functions can be applied to a correlation matrix, their use is somewhat cumbersome when applied to a correlation matrix. Instead, it is easier to use an older, more general function called `melt()` from the `reshape2` package (Wickham, 2007). Without the specification of any columns to not melt into the long form, the function takes all columns of the correlation matrix and melts (i.e., gathers) them into a data structure with one correlation per row, then outputs the result to the tidied data table, here *Rmelt*. The function assigns the `value.name` parameter, here set to `"cor"`, as the variable name for the resulting column correlation coefficients.

The created *Rmelt* data table contains three variables, the row label from the R matrix, the column label from the R matrix, and the correlation. There are 16x16 or 256 rows of data in the long form, of which an excerpt appears in Listing 5.5.

```
> Rmelt
      Var1 Var2    cor
1      m07  m07   1.00
2      m06  m07   0.52
3      m10  m07   0.40
4      m09  m07   0.32
       ...
255    m04  m03   0.07
256    m03  m03   1.00
```

Listing 5.5: Excerpt from the 256 rows of the long form data table, *Rmelt*.

The `ggplot()` function call in this example follows the usual pattern of the data specification, mapping the variables to the visual aesthetics, *Var1* and *Var2* in the long form data table, and the filled value that colors the heat map. From this information, the `geom_tile()` function generates the heat map, though several customizations modify the appearance of the visualization.

The primary consideration is the scaling of the plotted colors. By default, `ggplot2` produces a sequential scale from dark to light blue, though a divergent scale provides more differentiation, specified in the above code with `scale_fill_gradient2()`. The parameters `low`, `mid` and `high` specify the distribution of colors across the palette.

`scale_fill_gradient2()`
ggplot2 function,
Section 3.4.1, p. 64.

As indicated for the `lessR` plot in Figure 5.19a, when constrained to grayscale, a sequential palette is more appropriate since both sides of the divergent palette are the same color, gray. Better to vary continuously along a single dimension of dark to light or light to dark. This divergent function is shown in this example, however, to provide an example that can easily be converted to a more colorful palette without the constraint of grayscale. For example, set `low` to `"red"`, `mid` to `"gray80"`, and `high` to `"black"` to obtain more effective pattern recognition than that from Figure 5.19b.

`ggplot2` offers an option not provided by `lessR`, which is scaling the palette, here accomplished with the `limit()` function set to the extremes of the values of the correlation coefficients, `c(-1,1)`. Combined with the feature of `scale_fill_gradient2()` to provide a specified midpoint with `mid`, the colors in the heat map range from the darkest color of one color of the divergent palette for the value of -1, to neutral gray for 0, to the darkest color of the other color for a correlation of 1.

In contrast, the `lessR` heat map displays colors relative to the values in the correlation matrix, scaled from the lowest value to the highest value. The result is that this relative scaling more clearly delineates the patterns with more concentrated color variation. However, dropping scaling did not appreciably increase the range of color variation in the `ggplot2` visualization, even when run as an unconstrained, sequential grayscale as with the `lessR` visualization. The diagonal of the `ggplot2` heat map includes the correlation of 1, which absorbs much of the range of color variation since no correlations between two distinct variables approach this value.

To account for the remaining function calls in this `ggplot2` example, `coord_fixed()` has the default value of its `ratio` parameter set to 1, which contrains the two axes of the heat map to the same size, a square. The `labs` function removes the axis-labels for the two axes by setting each to `""`. By default, `ggplot2` does not rotate the value labels on the horizontal axis. So that they do not overlap each other, rotate them 90 degrees according to the `theme` function with the `axis.text.x` parameter set to the `element_text()` function with the `angle` parameter set to 90.

5.4 Scatterplots for Large Data Sets

To plot thousands of points, the over-plotting problem becomes severe as multiple points plot at or near the same coordinates. With enough plotted points the subsequent over-plotting transforms the scatterplot into a mass of solid color in regions of high density. Yet regions of low density suffer from the opposite problem, represented mostly by the background color interrupted by an occasional plotted point. An example appears in Figure 5.21 of 5000 points, the values from each variable sampled from a standard normal distribution.

Resolve this problem either by binning to obtain a histogram, or by smoothing to obtain a density plot. First, consider the smoothing option.

5.4.1 Smoothed Scatterplots

Figure 5.22 shows smoothed scatterplots of the 5000 points from Figure 5.21. By default, for more than 2500 rows of data to analyze, the `lessR` scatterplot function `Plot()` implicitly calls the base R function `smoothScatter()` to implement bivariate smoothing across the scatterplot, consistent with the current color theme.

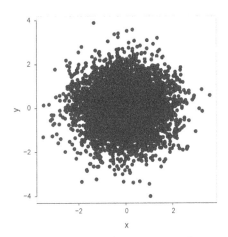

Figure 5.21: Plot of 5000 points from standardized normal distributions.

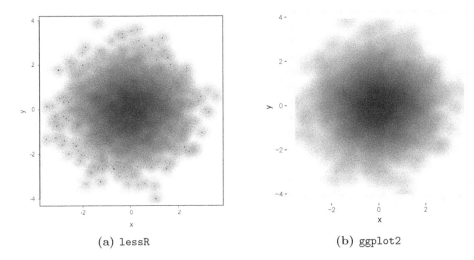

(a) `lessR`   (b) `ggplot2`

Figure 5.22: Scatterplot of smoothed, binned 5000 data values.

To turn smoothing off, set the parameter `smooth` to `FALSE`. Or, set to `TRUE` to obtain smoothing for less than 2500 points to plot. For ggplot2, `stat_density2d()` smooths the scatterplot with reference to the `"tile"` geom.

user's workspace,
Section 1.2.1, p. 7

Neither the data entered into a `lessR` analysis function nor in `ggplot2` need be in a data frame. Here define variables x and y in the R user workspace, more formally called the global environment. RStudio shows the contents of this workspace with the Environment tab, by default at the top-right corner of the RStudio window. `lessR` searches the user workspace for the relevant variable names before searching the default data frame. If the `data` parameter is specified, however, `lessR` goes right to the specified data frame. For `ggplot2`, specify a data frame as `NULL`.

R Input *Smoothed scatterplot from simulated data*

data: `x <- rnorm(5000); y <- rnorm(5000)`

lessR: `Plot(x, y)`

ggplot2: `ggplot(NULL, aes(x, y)) +`
`        stat_density2d(geom="tile", aes(fill=stat(density)^0.25),`
`          contour=FALSE) +`
`        scale_fill_gradientn(colours=c("gray40","gray80"))`

To visualize the smoothed scatterplot, map the 2-dimensional densities into colors. This function contains an exponent amenable to user manipulation. The `lessR` default is the base R default of 0.25, the value also specified for `ggplot2` in the above example. For `lessR`, override the default with the parameter `smooth.exp`. For `ggplot2` change the value of the exponent 0.25 in the above example. Larger values of the exponent lead to less color away from the most dense regions and smaller values lead to more color.

The 2-D densities are computed from a 2-D grid defined along both axes. A density is computed for each block, a 2-D bin, within the grid. The default for `Plot()` retains the `smoothScatter()` default of 125 equally spaced grid points along each axis, manipulated with the parameter `smooth_bins`. For `stat_density2d` the default is 100 grid points, manipulated with the parameter `n`. As with the construction of histograms, more data provides for a finer grid, though the default values appear to work well in most situations. Especially with 2-D density estimation, CPU time increases considerably for larger numbers of grid points.

By default, `Plot()` also plots the first 100 points with the least density to facilitate outlier identification. Control the number of such plotted points with `smooth.points`. Control the size of the plotted points with `smooth.size`.

5.4.2   Contoured and Hex-Binned Scatterplots

As with the analogy in 1-dimension with a histogram vs. a density plot, the smoothed density plot generally seems preferable to the discrete histogram in the representation of continuous data. As an alternative, however, `ggplot2` offers both a contoured version of the scatterplot and a 2-D binned version of the scatterplot, shown in Figure 5.23.

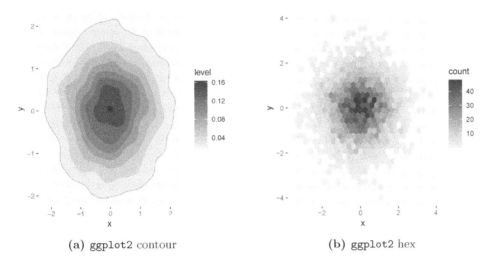

(a) ggplot2 contour   (b) ggplot2 hex

Figure 5.23: Countered and hex scatterplot of 5000 data values.

Obtain the contour plot with `stat_density_2d()`, specifying `"polygon"` for the `geom` and filling the contours with the calculated value of `level`. As with previous examples, the `stat` function specifies calculated variables in the visualization.

Accomplish the 2-dimensional binning with `geom_binhex()`, which is all that is required after the data source is specified and the variables to plot with `aes()`

R Input *Contoured and hex-binned scatterplots from simulated data*

data: `x <- rnorm(5000); y <- rnorm(5000)`

---

ggplot2: `ggplot(NULL, aes(x, y)) +`
        `stat_density_2d(geom="polygon", aes(fill=stat(level))) +`
        `scale_fill_gradientn(colors=c("gray80","gray20"))`
ggplot2: `ggplot(data.frame(x,y, aes(x,y)) + geom_hex() +`
        `scale_fill_gradientn(colours=c("gray80","gray20"))`

Both the contour and the hex plots present continuous data as discrete, so the smoothing option can more accurately portray the joint distribution. Some analyses, however, may require information from the levels of the contours, or the counts in each bin provided by `geom_hex()`. Each alternative conveys a different type of potentially useful information.

Chapter 6

Visualize Multiple Categorical Variables

6.1   Two Categorical Variables

The two-variable bar chart provides the most common visualization of the relationship between two categorical variables. This section explores three expressions of this bar chart: Stacked, unstacked, and Trellis. Also introduced is the related Trellis Cleveland dot plot.

bar charts of one variable,
Section 3.1.1, p. 47

Much of the `lessR` and `ggplot2` syntax and appearance for the two-categorical variable bar chart generalizes from the previously presented one-variable bar chart. As with the single categorical variable bar chart, a categorical variable x displays its levels on the corresponding axis. The corresponding values of the numerical variable y map to the height of each bar. For the two-categorical variable bar chart, the second categorical variable either divides each bar into separate rectangles, into different bars, or into separate panels, respectively corresponding to stacked, unstacked, and Trellis bar charts.

map data values into a visual aesthetic,
Section 10.2.1, p. 216

With `lessR` the default first two colors in the bars follow from the default qualitative palette, `"colors"`, which begins with a version of blue followed by brown. Use `style("gray")` to plot subsequent visualizations with colors implicitly chosen from the `lessR` `"grays"` palette. The default `ggplot2` behavior displays the colors for the two-variable bar chart as the first two colors from its default qualitative palette, a variation of orange and then aqua. Obtain grayscale, such as with the following `scale_fill_grey()` function call added to each of the following examples.

---
grayscale bar colors
ggplot2: `scale_fill_grey(start=.6, end=.3)`
---

vectors, Listing 1.3, p. 14

Customize the bar colors with the `fill` parameter for both `lessR` and `ggplot2`. Submit each set of custom colors as an R vector with the `c()` function.

6.1.1   Stacked Two-Variable Bar Chart

Figure 6.1 shows the default (subject to grayscale) `lessR` and `ggplot2` two-variable bar charts that portray the relation between *Dept* employed and *Gender*. The bar colors indicate the corresponding proportions of the `by` or `fill` variable for each level of the `x` variable. In Figure 6.1 each bar displays the proportion of Males and Females for the corresponding Department.

factor variables,
Section 1.2.6, p. 19

By default, R alphabetizes the presentation of the categories, Female before Male. To override the default ordering for the levels of any categorical variable, convert the variable to a factor variable with a specified custom ordering of the levels.

by parameter `lessR`,
Section 5.2.1, p. 113

To create the two-variable bar chart, for `lessR` again employ the `by` parameter that plots information from separate groups on the same panel. For `ggplot2`, define the second categorical variable as the value of the `fill` parameter for the `aes` function.

R Input *Default two-variable bar chart*
data: `d <- Read("http://lessRstats.com/data/employee.csv")`

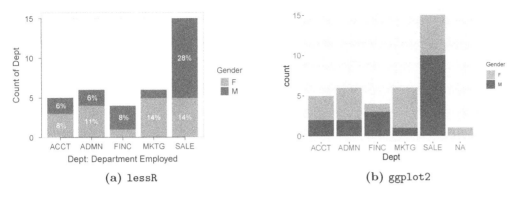

(a) lessR   (b) ggplot2

Figure 6.1: Default two-variable bar chart.

---

lessR: BarChart(Dept, by=Gender)

ggplot2: ggplot(d, aes(Dept, fill=Gender)) + geom_bar()

---

variable labels,
Section 1.2.5, p. 18

BarChart() provides the statistical analysis directed shown in Listing 6.1. The output begins with the variable names and, if present, variable labels. The primary output is the key concept for the analysis of the relation of two categorical variables: *joint frequency*, the frequency of occurrence for a combination of two levels of each of two categorical variables.

joint frequency: How often the same combination of values occurs on each of two categorical variables.

```
Dept: Department Employed
  - by levels of -
Gender: Male or Female

Joint and Marginal Frequencies
------------------------------

        Dept
Gender  ACCT ADMN FINC MKTG SALE Sum
   F       3    4    1    5    5  18
   M       2    2    3    1   10  18
 Sum       5    6    4    6   15  36
```

Listing 6.1: BarChart() two-way cross-tabulation table.

Construct the bar chart from this tabulation of the joint occurrences of the levels of the two variables across all combinations of levels, the *cross-tabulation table*. For example, Listing 6.1 shows that the company employs three female accountants and two male accountants. Either have the visualization function compute these values, or supply them directly.

cross-tabulation table: Table of joint frequencies for all combinations of levels of two or more categorical variables.

In addition to the joint frequencies, BarChart() provides the frequency distribution of each variable by itself, obtained for each level of one variable summed across the levels of the other variable. Refer to each sum in this context as a *marginal frequency*. These marginal frequencies appear under the Row or Column labeled Sum. Find the *grand total*, the sum of all joint frequencies, or, equivalently, the sum of the marginal frequencies for either one of the variables. From Listing 6.1, the grand total equals 36.

marginal frequency: Row or column sum from the table of joint frequencies.

grand total: Total sample size.

As shown in Listing 6.2, BarChart() also provides two statistics to evaluate the relationship between the variables.

```
Cramer's V: 0.415

Chi-square Test:  Chisq = 6.200, df = 4, p-value = 0.185
>>> Low cell expected frequencies, chi-squared approximation may not be accurate
```

Listing 6.2: `BarChart()` statistical output to evaluate the relationship among two categorical variables.

Cramer's V: Type of correlation for two nominal (categorical) variables that varies from -1 to 1.

p-value: Probability of obtaining a sample result as far from the null value of 0 given the population assumption of 0.

Cramer's V indicates the magnitude of association between nominal variables, the Pearson chi-square statistic rescaled with values between 0 and 1. The corresponding chi-square test is of the null hypothesis of no relationship between the two variables. Here the null hypothesis could not be rejected because the corresponding *p-value* exceeds the traditional $\alpha = 0.05$ threshold.

The stacked form compares the relative proportion of each level of the second categorical variable within each level of the first categorical variable. Base R, `lessR`, and `ggplot2` all implement the stacked bar chart as the default. Other versions of the two categorical variable bar chart, however, are available.

6.1.2   Unstacked Two-Variable Bar Chart

The unstacked or grouped bar chart portrays the levels of the second categorical variable as separate, adjacent bars. As Figure 6.2 illustrates, this form directly compares the second categorical variable to each other for each level of the first variable.

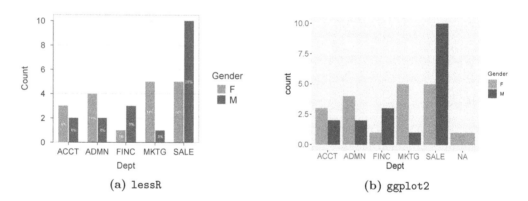

(a) `lessR`   (b) `ggplot2`

Figure 6.2: Unstacked two-variable bar chart.

To obtain the unstacked format with `lessR`, set the base R parameter `beside` to `TRUE`. For `ggplot2`, within the call to `geom_bar()` set the parameter `position` to `"dodge"`.

R Input *Unstacked two-variable bar chart*
data: d <- Read("http://lessRstats.com/data/employee.csv")

lessR: BarChart(Dept, by=Gender, beside=TRUE)
ggplot2: ggplot(d, aes(Dept, fill=Gender)) + geom_bar(position="dodge")

The bars are necessarily narrower for the unstacked version of the bar chart, so the font size of the value displayed on each bar for the `lessR` chart is smaller than for the stacked version. If too small to read for the size of the visualization, set `values` to `"off"`.

6.1.3 Trellis Plots or Facets

Developed and named by Cleveland (1993), a Trellis plot displays a visualization on a separate panel for each level of a categorical variable. With two levels of *Gender*, the Trellis plot of *Dept* by *Gender* consists of two separate panels, one for level F and one for level M.

Trellis plot,
Figure 4.18, p. 98

Trellis bar chart

Figure 6.3 displays a bar chart of the count of the data values for *Dept* separately for Males and Females, labeled as M and F. To obtain more descriptive labels than shown here, convert *Gender* from a character variable as read to a factor variable.

factor variable,
Chapter 1.2.6, p. 20

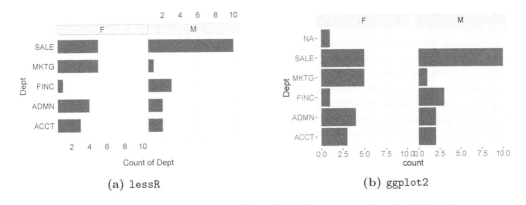

(a) `lessR`   (b) `ggplot2`

Figure 6.3: Trellis bar charts (facets) of *Dept* across levels of *Gender*.

With `lessR`, specify the second categorical variable with the `by1` parameter.[1] By default, `BarChart()` chooses the panel orientation according to the available space. If both vertical and horizontal orientation can be accommodated, the orientation is horizontal. Explicitly set the panel orientation with the `nrow` or `ncol` parameter. From the value of one of those two parameters `BarChart()` computes the value of the other.

With `ggplot2`, invoke the `facet_grid()` function with either the `rows` or `cols` parameter to display the facets as a horizontal grid or a vertical grid. This example refers to `cols` because the panels appear in separate columns. Then reference the `vars` function with the name of the categorical variable. To stack the panels horizontally, that is, in rows, specify the `rows` parameter instead of `cols`. Orient the bars horizontally with `coord_flip()` as a means to provide room to display the axis labels, the names of the categories.

---

[1] `lessR` Trellis plots implicitly constructed with the `lattice` (Sarkar, 2008) visualization package.

> **R Input** *Trellis (facet) bar charts*
>
> *data*: d <- Read("http://lessRstats.com/data/employee.csv")
>
> ---
>
> *lessR*: BarChart(Dept, by1=Gender)
>
> *ggplot2*: ggplot(d, aes(Dept)) + geom_bar() +
>   facet_grid(cols=vars(Gender)+ coord_flip()

To specify a second categorical variable from which to build a Trellis plot with `lessR`, invoke the `by2` parameter. With ggplot2, refer to both the `rows` and `cols` parameters within `facet_grid()`.

Trellis Cleveland Dot Plot

one-panel dot plot, Figure 3.3, p. 48

The Cleveland dot plot generalizes the one-panel visualization in Figure 3.3 to the multi-panel Trellis plot, such as shown in Figure 6.4. The second categorical variable in this example is *Gender*. Compare the number of men and women employed in each department, coded with the values of F and M.

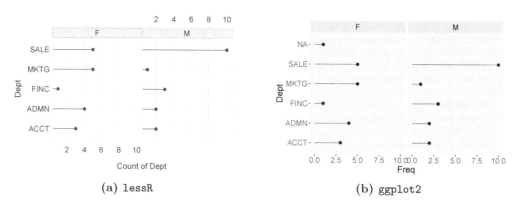

(a) `lessR`   (b) `ggplot2`

Figure 6.4: Trellis Cleveland dot plots of counts for *Dept*.

The `by1` parameter generates the `lessR` Trellis plot. Here set its value to *Gender* in the function call to `Plot()`. For ggplot2, first construct the summary data table with two-way counts for all combinations of *Dept* and *Gender* with `group_by()` followed by `summarize()`. Then pass the constructed data frame to `ggplot()`.

pipe operator, Section 2.5.1, p. 41

In this example, the last function in the chain of functions related by the pipe operator, %>%, is `ggplot()` itself, so do not explicitly identify the *d* data frame in the call to `ggplot()`. Then, to create the Trellis plot, add `facet_grid()` with the specification of the second categorical variable as `vars(Gender)`, displayed across columns.

> **R Input** *Trellis Cleveland dot plot of Dept counts by* Gender
>
> *data*: d <- Read("http://lessRstats.com/data/employee.csv")
>
> ---
>
> *lessR*: Plot(Dept, by1=Gender)
>
> *ggplot2*: d %>% group_by(Dept, Gender) %>% summarise(Freq=n()) %>%
>   ggplot() +

```
geom_point(aes(Dept, Freq), size=2.25) +
geom_segment(aes(x=Dept, y=0, xend=Dept, yend=Freq), size=.25) +
coord_flip() + facet_grid(cols=vars(Gender))
```

As with the previous example with `BarChart()`, generalize to a second Trellis variable with `Plot()` using the `by2` parameter. The given ggplot2 code generalizes to a second facet variable with both the `rows` and `cols` parameters specified within `facet_grid()`.

6.2 Other Styles for the Two-Variable Bar Chart

Within the three basic forms of a bar chart that relates two-categorical variables with a numerical variable – stacked, unstacked or grouped, and Trellis – there are many variations.

6.2.1 Sorted Two-Variable Bar Chart

Just as the values of the one-categorical variable bar chart can be sorted, the same applies to the two-variable format, as shown in Figure 6.5.

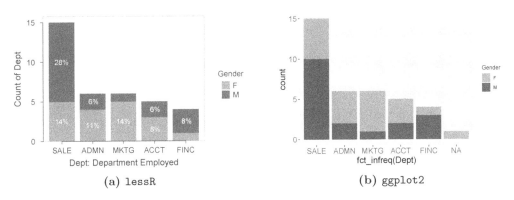

(a) `lessR`   (b) `ggplot2`

Figure 6.5: Sorted two-variable bar chart.

different strategy for ggplot2 sorting, Section 3.4.1, p. 62

As with sorting the values of the one-variable bar chart, with `lessR` invoke the `sort` parameter. Indicate an ascending or descending sort with a `"+"` or `-`, respectively. With `ggplot2`, in this example, use the `fct_infreq()` function, which reorders factor levels by frequency.

R Input *Descending sorted two-variable bar chart*

data: `d <- Read("http://lessRstats.com/data/employee.csv")`

lessR: `BarChart(Dept, by=Gender, sort="-")`
ggplot2: `ggplot(d, aes(x=fct_infreq(Dept), fill=Gender)) + geom_bar()`

Except for the `lessR` `sort` parameter and the `ggplot2` function `fct_infreq()`, the instructions to obtain the sorted version of the bar chart remains the same as for the default, unsorted version.

6.2.2   Horizontal Bar Chart

Figure 6.6 shows the horizontal version of the two-categorical variable bar chart.

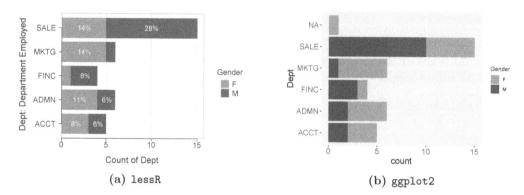

(a) `lessR`   (b) `ggplot2`

Figure 6.6: Horizontal bar chart.

horizontal bar chart,
Figure 3.2, p. 47

As with the one-categorical bar chart, with `lessR` set the base R parameter `horiz` to `FALSE`. For `ggplot2`, invoke `coord_flip()`.

> **R Input** *Horizontal bar chart*
> *data*: d <- Read("http://lessRstats.com/data/employee.csv")
>
> *lessR*: BarChart(Dept, by=Gender, horiz=TRUE)
> *ggplot2*: ggplot(d, aes(Dept, fill=Gender)) + geom_bar() + coord_flip()

Other than personal preference, an advantage of the horizontal orientation is the automatic accommodation of space for long value labels.

6.2.3   Bar Chart with Legend on the Top

By default, the visualization systems center the legend on the right-margin of the plot area. Figure 6.7 displays the legend on top of the bar chart.

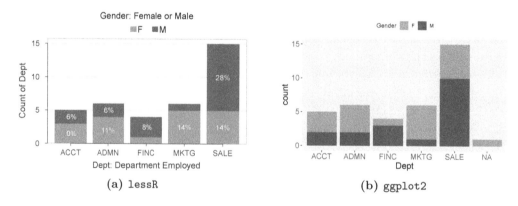

(a) `lessR`   (b) `ggplot2`

Figure 6.7: Two-variable bar chart with legend on the top.

Both `lessR` and `ggplot2` use the same parameter `legend_position` with value `"top"` to position the legend on the top. With `ggplot2` pass the parameter to `theme()`.

> **R Input** *Bar chart with legend on top*
>
> *data*: d <- Read("http://lessRstats.com/data/employee.csv")
> ___
> *lessR*: BarChart(Dept, by=Gender, legend_position="top")
> *ggplot2*: ggplot(d, aes(Dept, fill=Gender)) + geom_bar() +
>   theme(legend.position="top") +
>   scale_fill_grey(start=.6, end=.3)

Other values for `legend_position` for ggplot2 include `"none"`, `"left"`, `"right"`, and `"bottom"`. For more flexibility, specify a two-item vector in the x- and y-axes units for a more customized location. `lessR` relies upon base R `legend()`, which provides the following additional location values: `"bottomright"`, `"bottom"`, `"bottomleft"`, `"left"`, `"topleft"`, `"top"`, `"topright"`, `"right"`, and `"center"`.

6.2.4   100% Stacked Bar Chart

The stacked two-variable bar chart plots two categorical variables: The x-axis variable, plus the variable that determines the shaded areas on each bar, the `by` variable for `lessR` and the `fill` variable for `ggplot2`. The analysis here compares the percentage of the distribution of the `by` or `fill` variable *within* each level of the x-variable. What percentage of Men and Women are employed within each department? Answer this question with an analysis of the distribution of *Gender* within each level of *Dept*, shown in Figure 6.8.

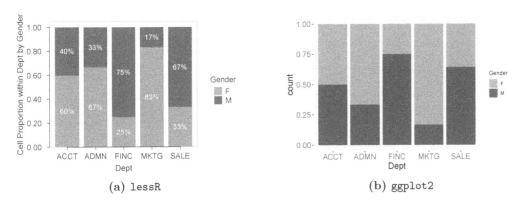

(a) `lessR`   (b) `ggplot2`

Figure 6.8: 100% stacked bar chart.

Each bar's height (length) for this analysis accounts for the full 100%, the full proportion of all elements within each group (category), x. Refer to the resulting visualization as the *100% stacked bar chart*. To create, for `lessR` set the `stack100` parameter to `TRUE`. For `ggplot2`, set the `position` parameter to `"fill"` within the call to `geom_bar()`.

100% stacked bar chart: Proportion of each level of the `by` or `fill` variable within each level of the x variable.

> **R Input** *100% stacked bar chart*
>
> *data*: d <- Read("http://lessRstats.com/data/employee.csv")
> ___
> *lessR*: BarChart(Dept, by=Gender, stack100=TRUE)
> *ggplot2*: ggplot(na.omit(d), aes(Dept, fill=Gender)) +
>   geom_bar(position="fill")

cross-tabulation table for two variables, Figure 6.1, p. 137

`BarChart()` provides the cross-tabulation table, shown in Listing 6.1, and adds the corresponding table of the cell proportions within each column, shown in Listing 6.3.

```
Cell Proportions within Each Column
------------------------------------

        Dept
Gender   ACCT  ADMN  FINC  MKTG  SALE
  F     0.600 0.667 0.250 0.833 0.333
  M     0.400 0.333 0.750 0.167 0.667
  Sum   1.000 1.000 1.000 1.000 1.000
```

Listing 6.3: `BarChart()` two-variable statistical analysis.

The numerical variable y is not limited to counts, as shown in the next section.

6.2.5 Bar Chart of Means across Two Categorical Variables

In addition to counts, another possibility defines a statistic such as the mean as the numerical variable, y. With `lessR` create this visualization of a statistic distributed across two categorical variables directly from the data, or from a summary table of the statistic across the two sets of categories. With `ggplot2`, more straightforward to compute a data frame of means from the raw data of measurements, then read the transformed data into `ggplot()`.

Bar chart from data directly

bar chart of means, Figure 3.11, p. 58

The bar chart of means across two categorical variables in Figure 6.9 appears in unstacked form.

beside parameter, Section 6.1.2, p. 138

To generalize from the bar chart of the means across the categories of a single categorical variable, as in Figure 3.11, include the `by` variable in the call to `BarChart()`. Specify the statistic to compute with the `stat` parameter. To obtain means, specify the value of `"mean"`. As indicated, set the `beside` parameter to `TRUE` to un-stack the bars.

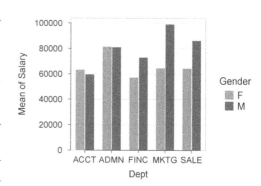

Figure 6.9: Bar chart of means for two variables direct from the data.

> **R Input** *Mean Salary for Dept-Gender combinations from Dulu*
> `d <- Read("http://lessRstats.com/data/employee.xlsx")`
>
> *lessR*: `BarChart(Dept, Salary, by=Gender, stat="mean", beside=TRUE)`

For `lessR`, follow the same syntax to create the bar chart of a statistic across two categorical variables for both the raw data of original measurements or aggregated data. `BarChart()` senses the form of the data, and either process the aggregate data directly, or computes the aggregate data from the raw data, the data of original measurements. In either case, if the data are not in the d data frame, then specify the name of the data frame with the `data` parameter.

Bar chart from aggregated data

As with the bar chart for a single categorical variable, for two categorical variables, the numeric variable can be entered as a summary statistic, here across the combinations of the levels of the two categorical variables. The resulting computed statistics organized as a data table are called *aggregate data*, a type of pivot table to use Excel nomenclature. To analyze these summary statistics for the two-variable bar chart, directly enter the coordinate data table from which the bar chart is constructed to the bar chart function.

To construct the coordinate data table, first read the original data table of measurements. In this example, to more descriptively label the output, as an option read the variable labels, the two-column table with each row defined by a variable name followed by the corresponding variable label. The *Gender* variable was coded with F and M. To display results with more descriptive value labels, convert the character string variable as read to a factor variable, such as with base R `factor()`.

Compute the summary table, the mean *Salary* for all combinations of levels of the five departments and, in this example, two genders, from the original data frame d with the `lessR` function `pivot()`. Previous examples used tidyverse functions `group_by()` followed by `summarise()` to accomplish the same result, though `pivot()` also provides the sample size `n` for each cell, as well as missing data information. Store the summary table in the data frame *d.mean*.

> **R Input** *Summary data table of Mean Salary across Dept-Gender Combinations*
>
> *lessR*: d <- Read("http://lessRstats.com/data/employee.xlsx")
> *lessR*: l <- Read("http://lessRstats.com/data/employee_lbl.xlsx",
>   var_labels=TRUE)
>
> ---
>
> *base R*: d$Gender <- factor(d$Gender, levels=c("F", "M"),
>   labels=c("Female", "Male"))
> *lessR*: d.mean <- pivot(d, mean, Salary, c(Dept, Gender))

Figure 6.10 displays the resulting summary table from within Excel. The data frame *d.mean* contains 10 rows, based on 36 observations. The original data frame of measurements, *d*, contains 37 rows of data, but the level of *Dept* is missing for one employee, so that observation is dropped from the aggregation.

Dept	Gender	n	Salary
ACCT	Female	3	$63,237.16
ADMN	Female	4	$81,434.00
FINC	Female	1	$57,139.90
MKTG	Female	5	$64,496.02
SALE	Female	5	$64,188.25
ACCT	Male	2	$59,626.20
ADMN	Male	2	$80,963.35
FINC	Male	3	$72,967.60
MKTG	Male	1	$99,062.66
SALE	Male	10	$86,150.97

Figure 6.10: Summary table of the means of *Salary*.

To export the *d.mean* data frame to MS Excel, here file *meanSalaries.xlsx*, use `lessR` `Write()` with the `format` parameter set to `"Excel"`, abbreviated `wrt_x()`. The row names in this data frame consist only of integers that indicate the row number. As such, do not write the row names to the file, so `Write()` by default sets the base R parameter

Margin notes

bar chart of means for one variable, Figure 3.11, p. 58

aggregate data: Statistics across combinations of the levels of one or more categorical variables.

coordinate data table to construct a bar chart, Figure 3.11, p. 58

variable labels, Section 1.2.5, p. 18

pivot() function, `lessR`: Split data into subsets from which to compute summary statistics.

Write() `lessR`, Section 1.2.7, p. 22

row.names to **FALSE**. Later, can re-read the Excel data table back into an **R** data frame of the same name with **Read()**.

R Input *Write and read data frame to and from an Excel file*
```
Write("meanSalaries", data=d.mean, format="Excel")
...
d.mean <- Read("meanSalaries.xlsx")
```

Calculated from the summary table, the coordinate data table, in Figure 6.10, Figure 6.11 presents stacked bar charts of mean *Salary* across the five departments and two genders.

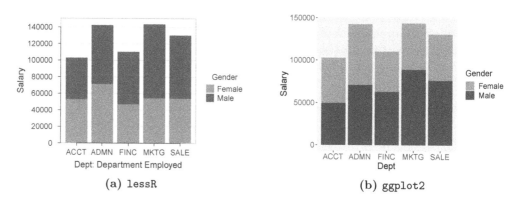

(a) `lessR`   (b) `ggplot2`

Figure 6.11: Two-variable bar chart direct from the coordinate data table.

R Input *Bar chart direct from data*
data: see above to obtain data frame d.mean

---

lessR: BarChart(Dept, Salary, by=Gender, data=d.mean)
ggplot2: ggplot(d.mean, aes(Dept, Salary, fill=Gender)) + geom_col()

As with most visualizations, `lessR` also presents output at the R console. The `BarChart()` output for this example appears in Listing 6.4. Examine the output to verify that the data values were properly read.

```
Data Table of Salary
--------------------

       Dept
Gender       ACCT         ADMN       FINC       MKTG        SALE
   F      53237.163   71434.003  47139.900  54496.022  54188.254
   M      49626.195   70963.345  62967.600  89062.660  76150.970
```

Listing 6.4: `BarChart()` two-variable direct from coordinate data table console output.

6.2.6 Two-Variable Cleveland Dot Plot

One of the primary analyses of the two-categorical variable bar chart compares the by variable across the levels of the x-variable. For example, in Figure 6.11 of mean *Salary* across combinations of levels of *Dept* and *Gender*, the proportion of the bar for *Gender* for the marketing department (MKTG) is larger for men than women, while for accounting (ACCT) the proportions appear approximately equal. This bar chart helps demonstrate the distribution by *Gender* across the various company departments. If such a comparison is of primary interest, however, then perhaps better to directly visualize the differences.

The two-variable Cleveland dot plot shows the dot plots of two different distributions across the levels of another variable, here of mean *Salary* across departments for men and women. Figure 6.12 illustrates this dot plot, with a line segment that connects the dots for each department to emphasize the comparison between values, here mean salaries. The default for the lessR visualization sorts by differences from the largest positive deviations to the largest negative deviations.

(a) lessR   (b) ggplot2

Figure 6.12: Two-variable Cleveland dot plot direct from data.

The coordinate data frame in Figure 6.10 of the group means is in tidy format, one value of mean *Salary* for each combination of levels of *Dept* and *Gender*, sometimes referred to as a long form data table. As of this writing, to generate the two-variable Cleveland dot plot with Plot(), the data table must be in wide form with the men's and women's mean *Salary* on the same line for each department.

To transform from the long form data table to the short form, here use the tidyverse tidyr function spread(). From Figure 6.10, the tidy version has three variables: *Dept*, *Gender*, and *Salary*. Consistent with all tidyverse functions, the first spread() parameter is the data source, here a data frame. The second variable, the key parameter, indicates the variable to "spread" into separate columns. Here spread the values of *Gender* into a column for Male and a column. The third parameter is value, the value to place in the body of the newly created columns. The result is Figure 6.13.

Transform a long-form (tidy) data table to wide form

```
tidyr: d.wide <- spread(d.mean, key=Gender, value=Salary)
```

The wide form of the data appears in Figure 6.13. The variable *Gender* no longer exists. Instead, separate columns appear for the values of *Gender* in this data set, Female and Male. Enter this wide form data table, *d.wide*, as the data for the lessR analysis. Enter the long form version, *d.mean*, as the data for ggplot2.

Dept	Female	Male
ACCT	$53,237.16	$49,626.19
ADMN	$71,434.00	$70,963.35
FINC	$47,139.90	$62,967.60
MKTG	$54,496.02	$89,062.66
SALE	$54,188.25	$76,150.97

Figure 6.13: The wide data table of means of *Salary*.

R Input *Two-variable Cleveland dot plot direct from data*
data: see above to get d.wide for lessR *and d.mean for* ggplot2

```
lessR: Plot(c(Female, Male), Dept, data=d.wide,
          ylab="Dept", xlab="Salary", legend.title="Gender")
ggplot2: ggplot(d.mean, aes(Salary, Dept)) +
          geom_line(aes(group = Dept), color="gray") +
          geom_point(aes(color = Gender))
```

6.2.7 Paired *t*-test Visualization

The paired *t*-test, or dependent groups *t*-test, analyzes the differences between matched pairs of data values. For example, obtain a pre-test score, then conduct an instructional session, which is then followed by a post-test. The analysis compares each test taker's pre-test score to the corresponding post-test score. Was there an improvement in scores on the post-test after the instructional session? If so, how much? The paired *t*-test provides the inferential test against the null hypothesis that the average population difference is zero.

A useful visualization for this analysis is the Cleveland two-variable dot plot that plots each set of paired scores direct from the original data table of measurements. To illustrate, consider the variables `Pre` and `Post` from the employee data set, scores before and after instruction. Figure 6.14 shows the corresponding Cleveland dot plot, again only for the first ten employees to save space.

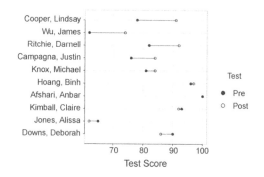

Figure 6.14: Cleveland dot plot of paired differences, data for the paired *t* test.

With lessR obtain this visualization in one of two ways: from Plot() directly as in the previous example, or from the ttest() function that implicitly calls Plot(). Creating the visualization from Plot() provides more user control, such as the ability to sort by differences and to customize the labeling of the axes and legend, as shown in Figure 6.14. The visualization, however, is automatic from ttest(), accompanied by a full paired dependent groups analysis.

R Input *Two lessR methods for two-variable Cleveland dot plot*

data: `d <- Read("http://lessRstats.com/data/employee.csv,row.names=1")`
    `d <- d[1:10,]`

---

lessR: `Plot(c(Pre, Post), row_names,`
        `xlab="Test Score", legend_title="Test")`

lessR: `ttest(Pre, Post, paired=TRUE)`

Indicate the paired or dependent-group *t*-test by invoking the base R parameter `paired`. In addition to the summary statistics, which here shows an average increase of 3.90 points on the post-test, `ttest()` console output includes the inferential analysis shown in Listing 6.5.

```
t-cutoff: tcut =   2.262
Standard Error of Mean: SE =   2.00

Hypothesized Value H0: mu = 0
Hypothesis Test of Mean:   t-value = 1.948,   df = 9,   p-value = 0.083

Margin of Error for 95% Confidence Level:   4.53
95% Confidence Interval for Mean:   -0.63 to 8.43
```

Listing 6.5: Dependent-groups `ttest()` statistical output.

Also included is the effect size, Cohen's `d`, to complement the significance test, shown in Listing 6.6.

```
Distance of sample mean from hypothesized:   3.90
Standardized Distance, Cohen's d:   0.62
```

Listing 6.6: Dependent-groups `ttest()` statistical output.

Figure 6.14 enhances the understanding of the results of the paired *t*-test. For example, in this analysis of *Pre* and *Post* scores, the *p*-value of 0.083 did not quite reach significance at $\alpha = 0.05$, though the effect size is reasonably large, $d = 0.62$. Examination of the Cleveland dot plot of differences reveals that although for most employees, scores on the post-test were higher, both Alisa Jones and Deborah Downs had higher scores on the pre-test, which likely lead to the lack of significance. Further, five of the ten employees did have higher scores on the post-test, and three had about the same score.

Why some employees improved, and others did not, would be worthy of more investigation. Perhaps the null hypothesis of no average difference between scores other than 0 is correct, so that any observed differences are sampling noise. However, given the small sample size, the reasonable effect size, and the low, albeit not significant, *p*-value, it is reasonable that the null hypothesis is false so that at least some employees learn from the instructional program.

6.3   Mosaic Plots and Association Plots

6.3.1   The Mosaic Plot

The *mosaic plot* (Hartigan & Kleiner, 1984; Friendly, 1994) provides a somewhat more elaborate visualization of categorical data than the bar chart, with the added benefit that it generalizes to more than two categorical variables. Construct the mosaic plot from a square, increasingly divided up into component rectangles according to the number of categorical variables plotted. The area of each rectangle, or tile, is proportional to the size of the corresponding joint frequency.

mosaic plot: A square subdivided into adjacent rectangular tiles, with the area of each tile proportional to the number of elements in that group.

Example: Department and Gender. To illustrate the construction of a mosaic plot, Figure 6.15a shows the visualization for a single categorical variable, *Dept*, from the employee data set. The plot divides the square into horizontal sections, each section corresponding to a level of *Dept*, its size proportional to the corresponding frequency. The two-categorical mosaic plot in Figure 6.15b illustrates the vertical slices through the square that correspond to the two values of a second categorical variable, *Gender*.

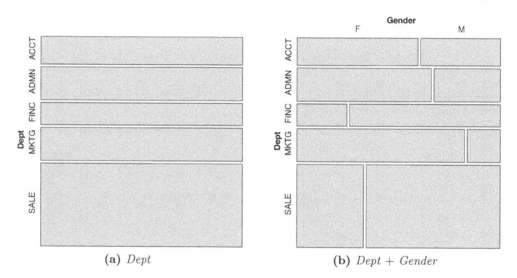

(a) *Dept*   (b) *Dept + Gender*

Figure 6.15: Mosaic plots of one and two categorical variables.

100% stacked bar chart, Figure 6.8, p. 143

Figure 6.8 introduced the 100% stacked two-categorical variable bar chart computed from the cross-tabulation table in Listing 6.1 for variables *Dept* and *Gender*, adjusted for the proportions within each column, as shown in Listing 6.3. The mosaic plot in Figure 6.15 presents an alternative representation of that bar chart, in this example, a horizontal version of the 100% stacked bar chart in Figure 6.8. There are five departments represented as rows, and two genders represented as columns. For the example of *Dept* with *Gender* illustrated in Figure 6.15, the number of Men in Sales has the largest joint frequency, 10, which corresponds to the largest rectangle in the mosaic plot, in the bottom right corner.

pie chart areas vs. circumference, Section 3.21, p. 71

For these data, for example, there are ten men in Sales and 5 women. So the tile that represents the joint frequency of Men in Sales, the largest tile, which is in the lower right-hand corner of Figure 6.15, has twice the length as the tile for Women in Sales. As with pie charts, in which areas can be compared, it is more efficient to compare lengths, along the circumference for pie charts and along an edge for the rectangular tiles of a mosaic chart.

Create this mosaic plot with the `mosaic()` function from the `vcd` package, an abbreviation for Visualize Categorical Data.

> **R Input**
> *data*: d <- Read("http://lessRstats.com/data/employee.csv")
>
> *vcd*: mosaic(~ Dept, data=d)
> *vcd*: mosaic(~ Dept + Gender, data=d)

R formula: An R expression that specifies a model.

Example: Survival on the Titanic. The previous example is of two categorical variables. This example of a mosaic plot, survival after the sinking of the Titanic, generalizes to three categorical variables, with one variable, *Survival*, explained in terms of *Class* of the passenger, and *Sex*.

This next example of the `mosaic` function develops an R *formula*, which provides a means for specifying a model, a functional relationship among variables. The expression of a model includes a variable of interest called the *response variable*, or outcome variable, or dependent variable. The model relates the response variable to one or more variables called explanatory variables or *predictor variables*.

response variable: The variable in a model that is explained by the remaining variables.

The previous example of *Dept* and *Gender* had no response variable, indicated by a function call that began with the tilde with no variable listed in front of it. This lack of a response variable implies that the analysis is not designed to explain the values of one variable in terms of others. Instead, analyze the joint frequencies, with all the variables of the same status. Place a variable in front of the tilde to define a response variable. The rectangles in the mosaic chart reflect the corresponding frequencies of this response variable at each combination of the explanatory variables.

predictor variable: A variable that predicts or explains the response variable.

This example is of the different classifications of the survivors of the RMS Titanic, the grand passenger ship that struck an iceberg and sank in the North Atlantic on its maiden voyage, April 14, 1912. The data, in the form of a four-way cross-tabulation table, are available as the data set called Titanic in the R `datasets` package, which is automatically loaded into memory when an R session starts. Table 6.1 presents the four classification variables and their associated levels.

No	Name	Levels
1	Class	1st, 2nd, 3rd, Crew
2	Sex	Male, Female
3	Age	Child, Adult
4	Survived	No, Yes

Table 6.1: Categorical variables and their values for the Titanic data.

Is *Class* of travel related to the chance of survival? How does the *Sex* (i.e., *Gender*) of each passenger and its relation to *Class* relate to survival? From the analysis of the *Titanic* cross-tabulation table, of the 470 females aboard the Titanic, 344 or

73.2% survived. Of the 1731 males aboard the Titanic, 367 or 21.2% survived. Of the 706 passengers in 3rd class, located at the bottom of the ship, 178 or 25.2% survived, and of the 325 passengers who made up 1st class, 203 or 62.5% survived. The corresponding Mosaic plot in Figure 6.16 visualizes these relationships. Dark gray represents survival and light gray indicates the opposite.

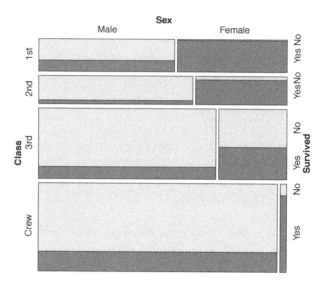

Figure 6.16: Grayscale mosaic chart for two categorical variables.

One characteristic of the data the visualization reveals follows from the length of the top edge of the boxes for men and women across the three classes. There were proportionally more women in 1st class than in 2nd class, and again, for 3rd class. This pattern necessarily reversed for the men. Almost all of the women in the first two classes survived, but not so for the women traveling in 3rd class where just more than half survived. Regardless of the class of travel, more men died than survived. Yet traveling 1st class was an advantage because the proportion of men who did not survive in 1st class was smaller than for the other classes. Only a small percentage of the crew were women, but unlike their much more numerous male counterparts, most survived.

Again use `mosaic()` function from the `vcd` package. The data are in the form of the cross-classification table, here *Titanic*. Express *Survived* as the response variable explained in terms of `Class` and `Sex`. By default, the colors are a very dark gray and a very light gray, optionally moderated here somewhat with the `highlighting_fill` parameter to a somewhat darker lighter color and a somewhat lighter darker color.

R Input *Analysis of Titanic survival with a mosaic plot*

data: `Titanic` four-way cross-tabulation table

```
vcd: mosaic(Survived ~ Class + Sex,
            highlighting_fill=c("gray85","gray45"), data=Titanic)
```

The mosaic plot generalizes the bar chart beyond two categorical variables, but it also can convey more information for any number of variables, explored next.

6.3.2   Independence and Pearson Residuals

The mosaic chart provides more additional information than does the corresponding bar chart: A visualization of the discrepancies between the obtained count on which the tile is constructed and the value expected if the categorical variables are independent, that is, not related. When two variables are independent, the probability of a response at one level of a categorical variable, is the same at all levels of the other variable. Said another way, two independent variables are not related in the sense that knowing the response to one variable provides no information about the response to the other. In this situation, the corresponding conditional probability is zero.

independent variables: The occurrence of the value of one variable provides no information regarding the value of the other variable.

For example, knowing a person's height provides no information as to their score on their recent statistics final. However, knowing a college student is tall increases the likelihood of that person playing college basketball. For the data analyzed here, is there a relationship between *Dept* and *Gender*? Is *Gender* distributed equally among all the departments? Or, do some departments tend to have higher proportions of Males or Females?

The null hypothesis of the test of independence is that the variables are independent, not related. The test, as any inferential test, is based on comparing the observed values, here the frequencies of each cell or cross-classification, with the values expected under the null. What is the expected frequency of a cell in cross-tabulation table for cell i,j given no relationship between the variables of interest? Represent the sum of row i with n_{+i}, the sum of column j with n_{j+}, and the grand total of all the observations with n.

$$\text{Expected Cell Frequency: } e_{i,j} = \frac{(n_{+i})(n_{j+})}{n}$$

The comparison of what occurred to what is expected is based on their difference, scaled by the square root of the expected, Pearson residual.

$$\text{Pearson Residual: } r_{i,j} = \frac{o_{i,j} - e_{i,j}}{\sqrt{e_{i,j}}}$$

The Pearson residual is scaled because the sum of all the squared Person residuals is the chi-square statistic, χ^2, upon which the inferential analysis follows.

The mosaic chart in Figure 6.17 provides more information than the related 100% stacked bar chart because it relates the tiles in the mosaic plot to the Pearson residuals. First, for these grayscale mosaic plots, tiles with positive residuals, a larger count than expected, are enclosed in solid lines and tiles with negative residuals, less counts than expected, are bordered with dotted lines. Without constrained to grayscale, the positive and negative residuals display in different colors if large enough. Also, the size of these residuals is then scaled by the corresponding legend.

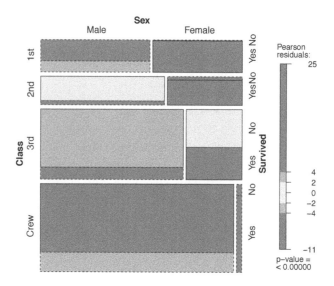

Figure 6.17: Grayscale mosaic chart for survival on the Titanic with residual-based shading.

As seen from Figure 6.17, the Pearson residuals are not large, all contained in the range from 1.2 to -1.2. In this particular two-categorical example, there were no residuals large enough to warrant the corresponding tile displayed in a darker shade of gray, in general, a darker version of the hue. This lack of large residuals is related to the lack of significance of the corresponding chi-square test, with a p-value larger than $\alpha = 0.05$ at 0.1847. For these data, there is no evidence that *Dept* and *Gender* are related, so the residuals are not interpretable beyond random noise. However, not only is the sample size very small for this small demonstration data set, $n = 36$, some individual joint frequencies are also below 5, which invalidates the chi-square statistic as a basis for testing the null hypothesis.

To implement the formula for this analysis, place the variable *Survived* to the left of the tilde. *Class* and *Sex* are the explanatory variables. The resulting mosaic plot shades each cell either according to the percent who survived or those who did not. Use a moderately dark gray value of `"gray40"` chosen to represent those who did survive and a lighter gray of `"gray85"` for those who did not. The values of the variable *Survived*, No and Yes, are ordered alphabetically, so the first color listed is for those who did not survive.

R Input *Mosaic plot with shading for Pearson residuals*
data: `Titanic` four-way cross-tabulation table

```
vcd: mosaic(Survived ~ Class + Sex, data=Titanic,
            gp=shading_hcl(Titanic, c=0, lty=1:2)))
```

The previous analyses of the Pearson residuals follow from the model of independence, from which the expected frequency for each cell is computed. Any model can be tested, however, if the analyst provides a data structure of the frequencies expected by a specific model of the same dimension and type as the data structure entered into the call to `mosaic()` with the parameter `expected`.

6.3.3   The Association Plot

Related to the mosaic chart is the association plot (A. Cohen, 1980; Meyer, Zeileis, & Hornik, 2003), which also consists of a set of rectangles (tiles). Each cell from the cross-tablation table is represented by a rectangle, which conveys much information regarding the expected and observed cell frequencies. The height of the rectangle is proportional to the cell residual. The width of the rectangle is proportional to expected value of the cell. As a consequence, the area of the box is proportional to the difference in observed and expected frequencies.

Find the association chart in Figure 6.18 that corresponds to the mosaic chart in Figure 6.15. The tiles in each row orient around a horizontal line. Tiles above the line indicate positive residuals and tiles below the line indicate negative residuals. Given the significant chi-square statistic, interpret the residuals in terms of the relationship between *Class* and *Sex* with *Survived*.

Again, without resorting to grayscale, different hues distinguish sufficiently large positive and negative residuals, by default blue and red, respectively. The horizontal baselines in Figure 6.18 separate positive and negative residuals, but the dotted lines around tiles below the baseline were retained for emphasis.

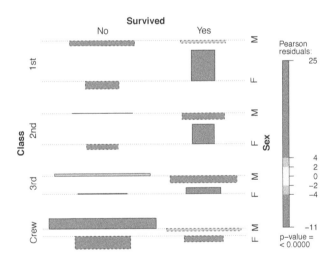

Figure 6.18: Grayscale association chart for Survival in terms of *Class* and *Sex*.

The highest tile in the association plot in Figure 6.18 is for women in 1st class who survived. The tile is above the horizontal base line, which indicates a positive residual, so proportionally more women in 1st class survived than expected by the independence model of no relation. The next highest tile indicates large positive Pearson residuals for women in 2nd class who survived.

Create the association plot with the `assoc()` function from the `vcd` package. Its functional form is the same as the related `mosaic()` function from the same package. Just change the name of the function.

R Input *Association plot with shading for Pearson residuals*
data: *Titanic* `four-way cross-tabulation table`

```
vcd: assoc(Survived ~ Class + Sex, data=Titanic,
           gp=shading_hcl(Titanic, c=0, lty=1:2)))
```

From a significant relationship between categorical variables according to the chi-square test, the mosaic and association plots both provide useful visual summaries as to the specific aspects of the relationship that likely contribute to the overall significance.

Chapter 7

Visualize over Time

process: Set of inter-related activities that generate output over time from given inputs.

A *process* generates its output over time. Examples include specific, well-structured processes such as filling each cereal box on the production line. A more general, complex process is a person growing from birth through adulthood, with output including such variables as height. Visualizing the process output over time contributes much to understanding the process output.

Time-oriented visualizations include time as a variable. One possibility expresses time in units of time, such as calendar dates, or seconds. Alternatively, represent time as an ordered sequence — first, second, etc. — without reference to the specific date or time of collection. First, consider the analysis of process output as an ordered sequence.

7.1   Run Chart and Control Chart

7.1.1   Run Chart

run chart: Scatterplot of ordered data values with each pair of adjacent points connected by a line segment.

A *run chart* plots the value of the numerical variable on usually the *y*-axis. For a given data value, plot the corresponding ordinal position according to time on the *x*-axis. A traditional name for the ordinal position variable is *Index*.

To create the run chart, plot the values of the variable of interest in their sequential order. An example appears in Figure 7.1 of an assembly line process that fills a cereal box. To calibrate the machine that releases the cereal into the box, record the weight in grams for 50 consecutive cereal boxes. The target value, the stated filled weight in grams on the cereal box, is 350 grams.

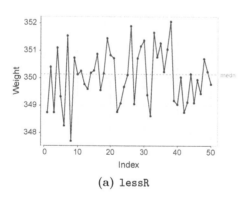 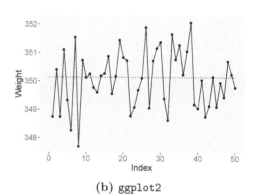

(a) `lessR`   (b) `ggplot2`

Figure 7.1: Run charts.

stable process: All process output generated from the same population.

Data values indicate a *stable process* if all are sampled from the same population, such as those in Figure 7.1. All data values of a stable process share the same underlying mean, standard deviation, and other characteristics. Their values differ only because of random sampling variation.

system in control: A stable process.

In the quality control literature, refer to a stable process as a *system in control*. One of the key implications of the evaluation of process stability is that only a stable process should be adjusted to improve accuracy or reduce variability. Accordingly, the analysis of stability is central to the field of quality control.

The run chart provides insight into the dynamics of an ongoing process. Because of the random variation inherent in every process outcome, the run chart displays these random fluctuations, including those data values that indicate a stable process. The visualization of the output of a stable process centers on a horizontal line with random variations of not too deviant values about that center line.

When do the data values not represent a stable process? As an example, consider a machining process.

- A single value could be sampled from another process, such as one machined part produced from a misaligned tool.

- At some point there could be a sudden shift of the mean. A machine cutter could shift and then lock into a different position, so that all subsequent values are sampled from a different process than the one that generated the first set of values.

- The mean could be continually changing over time, such as machine tool gradually shifting position, resulting in a trend. All data values from a continually shifting process are sampled from different processes.

To evaluate the stability of the underlying process, distinguish between two different sources of random variability. One source follows from the inherent sampling error of any sampling process about a stable set of population values such as the mean. This source of variability results in randomly variation about the mean.

Inadvertent sampling from other processes introduces another source of variability. Other processes have different means and standard deviations from the process of interest, so the corresponding points on the run chart generally break the pattern of the plotted points randomly varying about the mean of the process of interest.

The following sets of `lessR` and `ggplot2` code generate the run charts in Figure 7.1.

```
R Input Run chart.
data: d <- Read("http://lessRstats.com/data/Cereal.xlsx")

lessR: Plot(Weight, run=TRUE)
ggplot2: Index <- 1:length(d$Weight)
        md <- median(d$Weight, na.rm=TRUE)
        ggplot(d, aes(Index, Weight)) + geom_point() + geom_line() +
          geom_segment(aes(x=0, y=md, xend=length(Weight), yend=md),
                       linetype="dotted")
```

The `lessR` scatterplot function `Plot()` plots a run chart with the parameter `run` set to `TRUE` (instead of a VBS chart, the default for a numeric variable). By default, `Plot()` adds the median horizontal center line. `Plot()` labels the horizontal axis *Index* with sequentially ordered integer values.

VBS chart,
Section 4.5.1, p. 95

For `ggplot2`, separately create the *Index* variable, the integers ordered from 1 to the number of values. Also, compute the median for the center line. Specify the `geom_point()` function to plot the points, and `geom_line()` to plot the connecting line segments. Plot the center line with `geom_segment()`, which plots the center line as a line segment at the specified beginning and ending coordinates.

run: Sequence of consecutive points that all lie on the same side of the center line.

As with other `lessR` analyses, output includes a statistical analysis to supplement the visualization. The name "run chart" follows from the analysis of runs, provided by `Plot()` Listing 7.1 shows the default output of the `lessR` run chart.

```
n: 50
missing: 0
median: 0.1388874

Total number of runs: 25
Total number of values that do not equal the median: 50
Total number of values ignored that equal the median: 0
```

Listing 7.1: `BarChart` statistical output.

Obtain a full version of the run analysis with the parameter `show_runs` set to `TRUE`. As shown in Listing 7.2, this output lists every run over size 1. For example, the first reported run is Run 5, which means that the first four data values each define a run of size 1. That is, each of the first four values alternate on one side of the median and then the other side.

```
------------
Run Analysis
------------

size=2  Run  5:     5     6
size=2  Run 11:    12    13
size=3  Run 12:    14    15    16
size=4  Run 14:    18    19    20    21
size=4  Run 15:    22    23    24    25
size=3  Run 18:    28    29    30
size=2  Run 19:    31    32
size=6  Run 20:    33    34    35    36    37    38
size=5  Run 21:    39    40    41    42    43
size=3  Run 23:    45    46    47
size=2  Run 24:    48    49
```

Listing 7.2: Optional `BarChart` statistical output, all runs of length two or larger.

Several default values of the `lessR` run chart can also be changed. Instead of its default value of `"median"`, specify the `center_line` parameter as `"mean"`, `"zero"`, or `"off"`. As is true of any of the scatterplots from `Plot()`, turn off the points, in this case with only the line segments remaining, by setting `size` to 0.

7.1.2   Control Chart

The *control chart* provides more guidance to detect underlying process instability than the run chart. The control chart for individual data values begins with the same visualization as the run chart. The control chart adds upper and lower values called control limits to assist in identifying unusual data values, as well detect runs that indicate an unusual pattern. Either violation indicates an unstable process, a system out of control in the language of quality control.

Many types of control charts exist. Ideally, the control chart samples multiple data values, typically at least five or six, at each time point. The control chart plots either the means of the data values at each time point, or the standard deviation of each group. Control charts also are available for qualitative data, such as for plotting proportions.

Here consider a simple control chart of the individual data values, basically an enhanced run chart. Figure 7.2 presents control charts from the qcc package, accessed earlier for the Pareto chart, and also the ggQC package.

Pareto chart from package qcc, Figure 4.19, p. 100

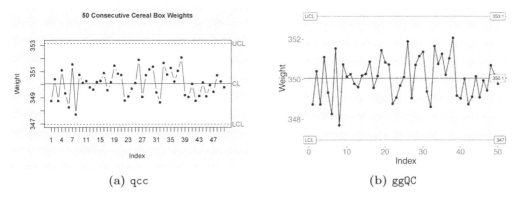

(a) qcc   (b) ggQC

Figure 7.2: Control charts of an apparently stable process.

The function from the qcc package is of the same name, qcc. Call by directly referencing the variable of interest, here *Weight*, located within the *d* data frame. The parameter type indicates the type of control chart, with the value "xbar.one" to plot individual data values. By default, the plot window contains various statistics regarding the data, here deactivated by setting add.stats to FALSE. Define the horizontal axis label as *Index* by the base R parameter xlab.

The ggQC package extends the functionality of ggplot2. Its individual functions are so well integrated with ggplot2 that the first steps to creating the visualization are to reference the standard ggplot() and aes() functions. Explicitly generate the values of the *Index* variable. Then generate the ggplot2 run chart as indicated in the previous example, followed by adding the upper and lower control limits with the stat_QC() function with the method parameter set to "XmR". This setting identifies a plot of the individual data values (X). The variability is estimated from the moving range (mr). Label the control limits and center line with the stat_QC_labels.

R Input *Control chart of a stable process.*

data: d <- Read("http://lessRstats.com/data/Cereal.xlsx")

```
qcc: qcc(d$Weight, type="xbar.one", add.stats=FALSE, xlab="Index",
        ylab="Weight", title="50 Consecutive Cereal Box Weights")
```

```
ggQC: Index <- 1:length(d$Weight)
       ggplot(d, aes(Index, Weight)) + geom_point() + geom_line() +
         stat_QC(method="XmR") + stat_QC_labels(method="XmR")
```

The two types of diagnostic tests to detect an unstable process are based on large
deviations from the center, and unusual patterns of runs, such as a run that is too
long, usually of length seven or more. The pattern of instability in Figure 7.3 is of a
large deviation, beyond the upper control limit, for the fifth data value. The qcc()
function enlargers the point size of a deviant plotted point, and also by default
displays the point in red, though here with a shade of gray. The same applies to
ggQC, except that it plots the deviant point as the same size as the remaining points.

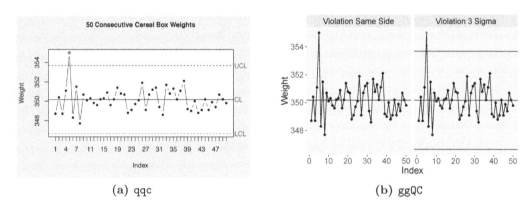

| (a) qqc | (b) ggQC |

Figure 7.3: Control charts of apparently an unstable process, that is, data values sampled
        from different populations.

To demonstrate the detection of a non-conforming data value, the fifth value in the
data file of the cereal package weights was arbitrarily increased. The result is the
spike in the control charts from the now revised data set.

The qcc() function flags potential violations of an underlying stable process, which
is unchanged from the previous example. The ggQC package detects instability
with the function stat_qc_violations(). By default, four distinct panels are
displayed, each isolating a potential source of instability. The two most important
such panels are the same sources of instability flagged by qcc(), large deviations,
and inconsistent patterns of runs. To request only these tests, set the show.facets
parameter to the vector c(1,4), as these two tests represent the first and fourth
panel. The function also allows customizing the colors of the control limit lines
and the center line with the rule.color parameter, and the color of any deviation
points with the violation_point.color parameter.

R Input *Control chart of an unstable process.*

data: d <- Read("http://lessRstats.com/data/Cereal.xlsx")
       d[5, "Weight"] <- 355

qcc: qcc(d$Weight, type="xbar.one", add.stats=FALSE, xlab="Index",
        ylab="Weight", title="50 Consecutive Cereal Box Weights")

```
ggQC: Index <- 1:nrow(d)
      d <- data.frame(Index, d$Weight)
      names(d) <- c("Index", "Weight")
      ggplot(d, aes(Index, Weight)) +
        stat_qc_violations(method="XmR", show.facets=c(1,4),
        rule.color="gray20", violation_point.color="gray50")
```

As indicated, many different types of control charts exist. Both `qcc()`, and `stat_QC()` with `stat_qc_violations()`, are capable of analysis of these different applications.

7.2 Time Series

Time series examples, Figure 2.5, p. 40

Data values plotted in their order of occurrence, labeled by their order, result in a run chart. For a time series, display the actual date and/or time for the generation of each data value.

time series visualization, Figure 2.10, p. 41

Chapter 2, the Quick Start chapter, provided three different forms of time series visualizations: one plot on one panel, multiple plots on the one panel, and multiple plots on different panels. The data for those examples are organized as tidy data, that is, in long form. Each row defines three values: The corresponding time period (month), the company, and the share price. Because the data were read from an Excel file with the date variable formatted as an Excel date type, R automatically converted the variable to an R date type.

7.2.1 Filled Area Time Series

Figure 7.4 extends the time series plots from Figure 2.10 by filling in the area under the curve and adding an annotation.

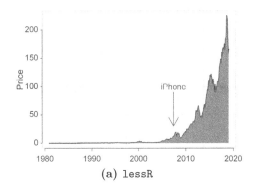
(a) `lessR`

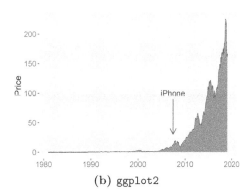
(b) `ggplot2`

Figure 7.4: Filled time series of Apple share price.

With `lessR`, plotting a variable of type `Date` as the *x*-variable in a scatterplot automatically creates a time series visualization. `Plot()` draws the connecting line segments, without the points at each time period (`size=0`). To add the area fill, for

filter parameter,
Section 2.10, p. 41

`lessR` set the `area` parameter to `TRUE` for the default color from the current color theme. Or, set to a specific color. Set the `rows` parameter to analyze only data for Apple.

For `ggplot2`, create a line chart with `geom_line()`. Add a call to `geom_area()` to fill in the area under the curve. The default fill is dark, lightened somewhat in Figure 7.4 by optionally setting the `fill` parameter to `"gray50"`. In this example, plot the time series only for Apple, that is, restrict the analysis to a subset of the rows of the original data frame, *d* with the tidyverse `filter()` function.

```
R Input  Annotated filled time series of monthly Apple share price
data: d <- Read("http://lessRstats.com/data/PPStechLong.csv")
      d$date <- as.Date(d$date, format="%Y-%m-%d")
      x <- as.Date("2007-06-01")

lessR: Plot(date, Price, rows=(Company == "Apple"), fill="on",
          add=c("iPhone", "arrow"), x1=c(x,x), y1=c(100,90), x2=x, y2=30)
ggplot2: d %>% filter(Company == "Apple") %>%
    ggplot(aes(date, Price)) + geom_line()+ geom_area(fill="gray50") +
      annotate("text", x=x, y=100, label="iPhone", color="gray20") +
      annotate("segment", x=x, xend=x, y=90, yend=30, color="gray20") +
      annotate("segment", x=x, xend=as.Date("2006-10-01"),
          y=30, yend=40, color="gray20") +
      annotate("segment", x=x, xend=as.Date("2008-02-01"),
          y=30, yend=40, color="gray20") +
      theme(axis.title.x=element_blank())
```

annotate,
Section 5.1.3, p. 109

Annotate the plot with the `lessR` parameter `add` for `Plot()`, and the `annotate()` function for `ggplot2`. Previously introduced for general scatterplots, for the time series express the *x*-coordinates for the annotations in terms of the `Date` type.

The annotations in Figure 7.4 consist of the text field `"iPhone"` with an arrowhead that points to the time that the first iPhone became available. With `lessR`, list each component of the annotation as a vector for `add`. Any value listed that is not a keyword such as `"rect"` or `"arrow"`, as shown in Table 5.1, is interpreted as a text field. Then, in order of their occurrence in the vector for `add`, list the needed coordinates for the objects, as specified in Table 5.1. To place the text field `"iPhone"` requires one coordinate, `<x1,y1>`. To place an `"arrow"` requires two coordinates, `<x1,y1>` and `<x2,y2>`. For example, the second element of the `y1` vector is the `y1` value for the `"arrow"`. The text field does not require a second coordinate, so specify `x2` and `y2` as single elements instead of vectors.

add fields, fields and
their required
coordinates,
Table 5.1, p. 111

With `ggplot2`, specify `annotate()` for the text field, line segment, and the two small segments for the arrowhead. What `lessR` refers to as `x1` and `x2`, `ggplot2` refers to as `x` and `xend`, and similarly for `y`. Specify the ends of the arrowheads as dates eight months removed on either side of the date for the introduction of the iPhone, indicated by the line segment.

7.2.2   Stacked Multiple Time Series

To compare the share prices over time for the three companies, the previously presented Figure 2.11 plots the time series for Apple, IBM, and Intel share prices on the same panel. Another comparison stacks the three plots, one on top of the other, illustrated in Figure 7.5. This form directly compares the share price at a given time by the size of the corresponding area under each curve, and also shows the aggregate value of all three share prices.

multiple time series on the same panel, Section 2.5.2, p. 43

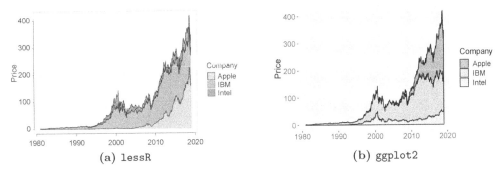

(a) `lessR`   (b) `ggplot2`

Figure 7.5: Stacked time series chart.

For the analysis of a long-form data table with `lessR`, plot multiple time series on the same panel with the `by` parameter. Set the value of `by` to the grouping variable, *Company* in this example. To create the stacked version of the visualization, set the parameter `stack` to `TRUE`. With `ggplot2`, set the parameter `fill` to the grouping variable in `aes()` to obtain multiple time series on the same panel. Invoke `geom_area()` to stack the separate time series plots.

multiple time series plot, Figure 2.11, p. 43

The meaning of the values on the time-axis, the *x*-axis, is self-explanatory, so `lessR` by default omits the axis label. With `ggplot2`, invoke `theme()` and re-set the `axis.title.x` parameter. Invoke grayscale fill with `scale_fill_grey()`.

R Input *Stacked time series chart*

```
data: d <- Read("http://lessRstats.com/data/PPStechLong.csv")
      d$date <- as.Date(d$date, format="%m/%d/%Y")
```

```
lessR: Plot(date, Price, by=Company, stack=TRUE, trans=0.4)
ggplot2: ggplot(d, aes(date, Price, fill=Company)) +
         geom_area(color="black", alpha=0.4) +
         theme(axis.title.x=element_blank()) +
         scale_fill_grey(start=0.1, end=0.7)
```

The order of the stacked layers follows from the order of the companies listed in the data frame: Apple, IBM, and Intel. `lessR` and `ggplot2` arbitrarily stack in opposite orders. Figure 7.5a lists Intel at the top of the stack whereas Figure 7.5b lists Intel at the bottom.

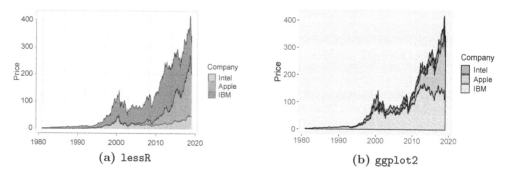

Figure 7.6: Stacked time series chart.

factors, Section 1.2.6,
p. 19

Figure 7.6 shows a stacking order of Intel, Apple and IBM. To customize the ordering, convert *Company*, read as type `character`, to type `factor`. Specify the levels in the desired order in the conversion to the variable as type `factor`. Here accomplish the re-ordering by specifying the new order of the rows of data with the base R `order()` function. Embed this function within the base R subsetting function, indicated with the square brackets.

> **R Input** *Reorder long-form data frame d to reorder stacking order*
>
> *data*: `d <- Read("http://lessRstats.com/data/PPStechWide.xlsx")`
>   `d$Company <- factor(d$Company, levels=c("Intel", "Apple", "IBM"))`
>   `d <- d[order(d$Company), ]`

Generate the stacked time series visualizations with the same `lessR` and `ggplot2` code from the previous example.

Both figures in Figure 7.5 have optional semi-transparent area fills. With `lessR`, set the `trans` parameter to 0.4. With `ggplot2` set the `alpha` parameter to the same value. Set the grayscale for `lessR` with `style("gray")`, which applies to all subsequent analyses. For `ggplot2`, set the line color to `"black"` with `color`. Set the fill area to shades of gray with `scale_fill_grey()`.

7.2.3   Formatted Multi-Panel Time Series

Trellis time series
visualization,
Figure 2.12, p. 44

Figure 2.12 illustrates the multi-panel time series plot, a Trellis plot, or a faceted plot as referenced by `ggplot2`. Figure 7.7 plots the same data, but formatted with a black background.

style() `lessR`,
Section 10.4.2, p. 227

To achieve this formatting, `lessR` provides parameter `sub_theme` as part of the call to `style()`. Set `sub_theme` to `"black"`, in conjunction with the `theme` of `"gray"`, to obtain another version of grayscale. The `sub_theme` resets many of the properties of the visualization consistent with this black background, such as a light gray font color for the various text labels. Also added was a transparency for the area fill under each time series plotted line. Setting `n.col` to 1 stacks the panels in a single column instead of their default row orientation.

theme() `ggplot2`,
Section 10.4.2, p. 227

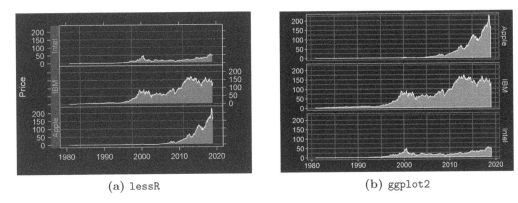

(a) `lessR` (b) `ggplot2`

Figure 7.7: Formatted Trellis (faceted) time series.

The `ggplot2 theme()` function customizes individual properties of the visualization. Obtain the full list from `get_theme()`, some of which are set in the following code.

R Input *Formatted Trellis (faceted) time series*

```
data: d <- Read("http://lessRstats.com/data/PPStechLong.csv")
      d$date <- as.Date(d$date, format="%m/%d/%Y")
```

```
lessR: style("gray", sub_theme="black", trans=.4)
      Plot(date, Price, by1=Company, fill="on")
ggplot2: ggplot(d, aes(date, Price)) +
      geom_area(fill="gray60", color="gray90", alpha=0.7) +
      facet_grid(rows=vars(Company)) +
      theme(plot.background = element_rect(fill="black"),
          panel.background = element_rect(fill="black", color="gray90"),
          panel.grid.minor = element_line(size=.5, color="gray25"),
          panel.grid.major = element_line(size=.5, color="gray25"),
          axis.text = element_text(color="gray90"),
          axis.ticks = element_line(color="gray90"),
          strip.background = element_rect(fill="black"),
          strip.text = element_text(color="gray90"),
          axis.title.x = element_blank())
```

With `style()`, `sub_theme` simplifies the formatting, yet the individual properties remain available for customization. Obtain the full list of `lessR` properties with `style()` by setting parameter `show` to `TRUE`. Here the visualizations are limited to grayscale, but in other contexts, choose any combination of fill area color and translucence, line color, and background and grid colors.

The following revisions to the above code create the Trellis time series plots that closely resemble the *Wall Street Journal* visualization style. The style consists of a translucent blue area fill with a brighter blue line color against a black background with white lettering.

Modify above code to achieve transparent blue fill in time series

lessR: `style(...  color="steelblue2", area.fill="steelblue3")`

ggplot2: `...  geom_area(color="steelblue2", fill="steelblue3", alpha=0.7)`

To obtain Figure 7.7 with blue shading instead of gray, for `lessR` add the `color` and `area.fill` parameters as specified in the call to `style()` with the `sub_theme` and `trans` parameters. Also, changing the color theme with `style()` changes the line and fill in the time series plot according to the chosen color theme. For `ggplot2`, change the values of the `color` and `fill` parameters of the `geom_area()` function.

7.2.4   Data Preparation for Date Variables

One possibility reads the dates from character strings as stored in the text or Excel data file, and then converts. Another possibility reads from an Excel file that represents the dates as the Excel date type. These two possibilities are discussed next.

Convert Character Strings to Dates

To create a time series with dates, specify the dates as a variable of type `Date`. Excel and R provide for variables typed according to the meaning of their data values. When the `lessR` and `tidyverse` read functions, `Read()` and `read_csv()`, read Excel files with a date field into a data frame, the conversion to R dates is automatic.

Text files, however, store all information as character strings. The read functions automatically convert numeric data values from character strings stored in a text file to numeric variables stored within the data frame. However, R does not provide an automatic conversion to a date type such as type Date when reading dates as character strings. Instead, by default, R reads the character string dates into a data frame as type `character`.

How to convert dates read as character strings into a data frame to an R date type such as `Date`? Invoke the base R function `as.Date()`[1]. The complication is that dates are expressed in many different formats. For example, represent the first day of July, 2015 as "July 1, 2015" or "7/1/2015" or "1-7-15". The key to using `as.Date()` is its `format` parameter, which defines each part of the date character string – year, month, and day – with a formatting code.

Each code begins with the % sign. The code %Y identifies a four-digit year in the character string. Identify a year expressed as two digits with %y. To identify an expression of the month as digits from 1 to 12, use %m. Identify a month spelled out alphabetically in full with %B, and an abbreviated alphabetic month with %b. Identify the day of a month with %d. Include any delimiters in the character string such as a hyphen, forward slash, blank, or comma in the `format` expression in the same position found in the data values.

---

[1]Another method converts the data values as they are read with the `colClasses` parameter on the read statement, which informs R as to the class of the variable to read.

R has a default date format, the ISO 8601 international standard: a four-digit year, a hyphen, a two-digit month, a hyphen, and then a two-digit day. For example, express the first day of July, 2015 in this standard as "2015-07-01". Dates expressed as character strings in this format, or with forward slashes instead of hyphens, need no explicit format parameter in the call to as.Date(). All other date representations embedded in a character string require the format parameter for R to transform to an R date.

strptime() base R, help file provides all the format codes (enter **?strptime**)

Listing 7.3 presents a variety of character string expressions of dates, all of which output to the same date, "2015-09-01".

> **R Input** *as.Date() function calls that transform different character strings to the same date: "2015-09-01"*
>
> ```
> as.Date("2015-09-01")
> as.Date("2015-09-01", format="%Y-%m-%d")
> as.Date("2015/9/1")
> as.Date("2015.09.01", format="%Y.%m.%d")
> as.Date("09/01/2015", format="%m/%d/%Y")
> as.Date("9/1/15", format="%m/%d/%y")
> as.Date("September 1, 2015", format="%B %d, %Y")
> as.Date("Sep 1, 2015", format="%b %d, %Y")
> as.Date("20150901", format="%Y%m%d")
> ```

Listing 7.3: Alternate expressions of dates in character strings that decode to "2015-09-01" with a call to as.Date() given the proper format.

To illustrate, read the data from the web file PPStechLong.csv with Read() into the data frame *d*, with the output shown in Listing 7.4.

```
> d <- Read(http://lessRstats.com/data/PPStechLong.csv)

    Variable                    Missing  Unique
      Name       Type   Values   Values  Values   First and last values
    -----------------------------------------------------------------------
  1        date character   1374        0     458   12/1/1980   1/1/1981 ...
  2   Company character   1374        0       3   Apple   Apple ... Intel
  3      Price    double   1374        0    1259   0.027   0.023 ... 46.634
    -----------------------------------------------------------------------
```

Listing 7.4: Output of Read().

The output from Read() in Listing 7.4 indicates that the dates have been read as type character. The sample values that Read() provides, such as 12/1/1980, show that the dates are formatted as number of the month, number of the day of the month, and then a four-digit year, delimited by forward slashes. From this information, convert to the Date type as shown in Listing 7.5. Then invoke the lessR function details(), with default data frame *d* and with abbreviation db() for the brief version, the same function that Read() relies upon to document the data frame.

details(), lessR function:
Section 1.2.4, p. 16

```
> d$date <- as.Date(d$date, format="%m/%d/%Y")
> db()

     Variable                  Missing  Unique
         Name     Type  Values  Values  Values  First and last values
     --------------------------------------------------------------------
  1      date     Date    1374       0     458  1980-12-01 ... 2019-01-01
  2   Company  character  1374       0       3  Apple  Apple ... Intel
  3     Price    double   1374       0    1259  0.027  0.023 ... 46.634
     --------------------------------------------------------------------
```

Listing 7.5: The variable *date* properly converted to a variable of type `Date`.

Once converted with `as.Date()`, as shown in Listing 7.5, the dates follow the R default ISO standard of a four-digit year, two-digit month, and two-digit month, each separated by a hyphen.

Read Excel Date Type Directly as Dates

serial number:
Integer that represents a date as the number of days from an origin, a beginning date.

Both R and Excel store data values as a date type internally as integers, called *serial numbers*, sequentially numbered from an arbitrarily set origin. R uses the origin `"1970-01-01"`, which corresponds to serial number 1. Windows Excel, and Macintosh Excel versions since 2011, standardize on the origin `"1900-01-01`. Versions of Macintosh Excel 2011 and earlier use `"1904-01-01"`.

If reading a file from Macintosh Excel older than 2011, best to check to make sure the dates are read correctly. Or, convert the Excel date fields to Excel text fields, then explicitly convert to dates with `as.Date()`. To covert to a text field within Excel, select the column of date fields, choose Data -> Text to Columns, then on the subsequent third window pane, choose `to Text`.

The function `as.Date()` also converts serial numbers to R `Date` fields. To convert, provide the serial number as an integer and then the corresponding origin. Listing 7.6 shows this conversion for the Excel origin and the R origin.

> **R Input** *Different integers (serial numbers) that each output "2015-09-01"*
> ```
> as.Date(42246, origin="1900-01-01")
> as.Date(16679, origin="1970-01-01")
> ```

Listing 7.6: Alternate expressions of dates as integers that decode to the "2015-09-01" with a call to `as.Date()` with the proper origin.

Wide and Long-Form Data Tables

data table,
Section 1.2.2, p. 7

time series data,
Figure 2.9, p. 41

Data analysis begins with a well-structured rectangular table of data. As applied to a time series, the data table example from Chapter 2, Figure 2.9 lists monthly share prices for Apple, IBM, and Intel. In that data table each row of data corresponds to a single price per share for a single company at a single time. The resulting data table is only three columns wide, and 1374 rows long, 458 monthly share prices for each of the three companies. The descriptive name for this organization of a data table is the *long form* also referred to as tidy data (Wickham, 2014).

<u>Wide form of a data table</u>. The complementary data table organization, the wide form, presents each row of data as the date followed by the share prices of each of the three companies, shown in Listing 7.7.

```
> d[1:4,]
        date Apple   IBM Intel
1 1980-12-01 0.027 2.051 0.212
2 1981-01-01 0.023 1.945 0.196
3 1981-02-01 0.021 1.941 0.185
4 1981-03-01 0.020 1.910 0.191
```

Listing 7.7: First four rows of alternative wide-form data table *d* that corresponds to the long form in Figure 2.9.

The variable Company, which is explicit in the long form, appears in the wide form as three separate vectors, one for each value of Company: Apple, IBM, and Intel. The wide-form data are not tidy data because each of the three measurements represents a *different* observation.

All `ggplot2` functions accept only long-form data tables. `lessR` is amenable to both long-form and wide-form data structures for time series analysis, as well as an R data structure designed especially for time series analysis, `ts`.

To plot multiple time-series, the `lessR` analysis of the wide-form data frame requires a vector of column names, one for each column of data. The following code generates the same multiple time series plot as from the long form in Figure 2.9.

R Input *Time series visualization in Figure 2.9 from wide-form data table*
data: d <- Read("http://lessRstats.com/data/PPStechWide.xlsx")

lessR: Plot(date, c(Apple, IBM, Intel))

<u>Convert from wide-form to long-form of a data table</u>. If the data are received in wide form, then `ggplot2` requires their transformation to long form. Transform Listing 7.7 to the long-form version in Figure 2.9.

melt, `reshape2`, an alternative to `pivot_longer()`, Section 5.3.2, p. 129

The most straightforward conversion from wide to long form is with `pivot_longer()`, from the `tidyr` package (Wickham & Henry, 2019). Define the columns in the wide form to gather into a single column, here, the companies Apple, IBM, and Intel to gather into a single variable named *Company*. Specify the name of the new variable in the created long-form table with the parameter `names_to`. Specify the corresponding value associated with each company at each date with `values_to`. Either list the variables to *not* gather into a single column, preceded by a minus sign, or list those to gather. This example specifies *-date*, but that expression could be replaced by the list of the remaining variables *Apple:Intel*, which are the columns of data to gather.

R Input *Wide-form data table to long-form*
data: d <- Read("http://lessRstats.com/data/PPStechWide.xlsx")

tidyr: dl <- pivot_longer(d, -date, names_to="Company",
                           values_to="Price")

The newly created grouping variable Company created by `pivot_longer()` in the long form has three values: Apple, IBM, Intel. The resulting data frame is in the form of a tidyverse tibble.

```
> dl[1:4,]
# A tibble: 4 x 3
  date          Company  Price
  <date>        <chr>    <dbl>
1 1980-12-01    Apple    0.027
2 1980-12-01    IBM      2.05
3 1980-12-01    Intel    0.212
4 1981-01-01    Apple    0.023
```

Listing 7.8: First four rows of long-form data frame version of *d* that corresponds to the wide form in Listing 7.7.

ggplot stacked time series plot.
Section 7.2.2, p. 165

With the long form of the data frame obtained, *dl*, the data are ready for entry into `ggplot2` functions for plotting. If standard R data frames are preferred, then apply the `data.frame()` function to the tibble output from `pivot_long()`.

7.3   Forecasts

A widely regarded R forecasting package is `forecast` (Hyndman & Khandakar, 2008) package[2] for time series analysis. Find more information in the online textbook (Hyndman & Athanasopoulos, 2018). Forecasting with the `forecast` package begins from an R object called a time series or `ts` object, a data container created specifically for time series data.

7.3.1   Time-Series Object

A time series data object, of class `ts`, pairs a sequence of numeric measurements with their corresponding dates. Specify the dates when creating the time series object with the function `ts()`, instead of from the values of a variable read from a data table. Create the time series object with the data sorted from oldest date to the most recent date. To create a univariate time series object, that is, for a single variable, apply `ts()` time-ordered values from a single column of a wide-form data table, such as for the share price data illustrated in the columns of Listing 7.7. Or for a long form of the data, such as illustrated in Figure 2.9, apply the `ts()` function to the desired level of the grouping variable, here `Company` of interest.

R Input *Create and plot a time series object, a.ts, to obtain Figure 2.10*
data: `d <- Read("Read("http://lessRstats.com/data/PPStechWide.xlsx")")`
      `Apple.ts <- ts(d$Apple, frequency=12, start=c(1980, 12))`

lessR: `Plot(Apple.ts)`
forecast: `autoplot(Apple.ts)`

---
[2]The `forecast` package continues to be maintained, but is ultimately designed to be replaced by its successor, the tidyverse based `fable` (O'Hara-Wild, 2019).

In this example, `ts()` creates a time series object for Apple share price from a column named Apple in the *d* data frame, the wide-form. The `frequency` parameter indicates the number of time points before the seasonal pattern repeats. In this example, the data are monthly. Monthly data varies from 1 to 12 and then repeats, so has a `frequency` of 12. Data collected only once a year has a `frequency` of 1, the default value. Quarterly data has a `frequency` of 4, and weekly data has a `frequency` of 52. The `frequency` of data collected hourly depends on when the pattern repeats. If the hours repeat daily, the `frequency` is 24.

frequency
parameter, `ts()`:
Number of
observations before
the seasonal pattern
repeats.

Indicate the starting time value of the first observation with a two-valued `start` parameter. Specify the time period for which the frequencies are assessed, here the starting year, and the interval number within that year, here the 12th month. The first observation in the data set is December of 1980. Given this information, any actual dates in the data file are superfluous and not referenced in operations involving the time series object.

With `lessR` plot the times series object `ts` directly with `Plot()`. Although `ggplot2` only accepts a long-form data table, it provides a function, `autoplot()`, that adapts to a data structure such as a `ts` object. With `forecast`, plot the time series accessing `ggplot2` with `autoplot()`. The result is Figure 2.10.

The univariate time series object generalizes to a multivariate time series. Here, starting from the wide-form data table, select the three columns of share prices and then transform to an R `matrix` object with the base R `as.matrix()` function. Apply `ts()` to the resulting matrix.

time series plot
univariate,
Figure 2.10, p. 41

> **R Input** *Plot a multiple time series object, a.ts, to obtain Figure 2.11*
>
> *data*: `d <- Read("Read("http://lessRstats.com/data/PPStechWide.xlsx")")`
> `    a <- as.matrix(d[,c("Apple","IBM","Intel")], nrow=nrow(d), ncol=3)`
> `    a.ts <- ts(a, frequency=12, start=c(1980, 12))`
>
> ---
>
> *lessR*: `Plot(a.ts)`
> *forecast*: `autoplot(a.ts)`

The results are the same visualizations in Figure 2.11, which present three time series on the same panel for `lessR` and `ggplot2`.

time series plot
multivariate, on the
same panel,
Figure 2.11, p. 43

7.3.2 Seasonal/Trend Decomposition

Much of data analysis is the search to uncover patterns blurred by the ubiquitous sampling instability. Flip a fair coin ten times and get six heads. Flip the same fair coin another ten times and maybe get four heads. There is underlying stability and pattern. The coin is fair so the probability of a head is 0.5. If it is not known, however, if the coin is fair, it is difficult to know this probability precisely without a large sample.

These fundamental concepts also apply to creating a forecast from time series data. Identify the underlying structure, then extend this pattern into the future to create a forecast. What is the structure that underlies a time series? Consider the three components shown in Table 7.1.

Pattern	Explanation
`trend`	Long-term, general, gradual increase or decrease of Y
`season`	Regular, relatively short-term repetitive up-and-down fluctuations of Y
`cycle`	Gradual, long-term, up-and-down potentially irregular swings of Y

Table 7.1: Three structural components of a time series of the variable Y.

The trend pattern may be linear or non-linear, characterized by a steady increase or decrease over the length of the time series, though usually with some minor, temporary dips in the opposite direction due to sampling error. For much business and economic data, measure the seasonal component in quarters of the year, such as the four seasons of Winter, Spring, Summer, and Fall. Measure the cyclical component in periods of many years, usually decades, and so is not present in time series analysis over lessor time periods. This component reflects broad, not necessarily symmetric swings about either side of the trend line.

Consider a time series of units of sales at a retail outlet of a product quarterly from 2013 through 2018. Figure 7.8 shows the time series data in the form of a `ts()` object. Figure 7.9 reveals that the time series exhibits increasing trend and seasonality.

```
> Sales.ts
       Qtr1   Qtr2   Qtr3   Qtr4
2013   63.82  94.52 134.09 120.54
2014  131.28 229.26 132.79 162.22
2015  202.98 165.03 193.73 213.43
2016  169.66 285.10 252.59 215.01
2017  229.13 240.96 342.32 284.07
2018  217.50 265.03 338.93 343.11
```

Figure 7.8: Sales data as a `ts` object.

Create the time series with `ts()`, and plot with `Plot()` or `autoplot()` as previously indicated. Because data are quarterly, the `frequency` is 4. The start date of the time series is the first quarter of 2013.

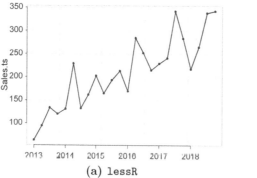

(a) **lessR**

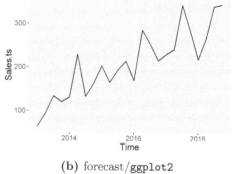

(b) forecast/**ggplot2**

Figure 7.9: Seasonal time series.

R Input *Seasonal time series*

```
data: d <- Read("http://lessRstats.com/data/Sales.xlsx")
      Sales.ts <- ts(d$Sales, start=c(2013,1), frequency=4)
```

lessR: `Plot(Sales.ts)`

forecast: `autoplot(Sales.ts)`

The specific combination of trend, cyclical, and seasonal components that defines the underlying pattern of a time series can only be revealed obscured by random error. With this deconstruction, apply the values of the components to the future time values to generate a forecast. Base the forecast on the underlying pattern delineated from the random error or noise that obscures this pattern.

The stable, predictable component of a time-series is the specific combination of trend, cyclical, and seasonal components that characterize the particular time series. How do these components combine to result in a value of Y at Time t, Y_t? One of two forms accounts for the underlying pattern, an additive model or a multiplicative model. The additive model expresses Y_t as the sum of the trend, cyclical, seasonal, and error components.

The cyclical, seasonal, and error components are each a fixed number of units above or below the underlying trend. For example, if the seasonal component at Time t is $S_t = -12.89$ and there is no cyclical component, then the effect of that season is to lower the value of Y_t the amount of 12.89 below the affect of the trend component. If the error component is 4.23 at Time t, then the effect of error on Y_t is an increase of 4.23 above the combined influence of trend and seasonality, illustrated in Figure 7.10.

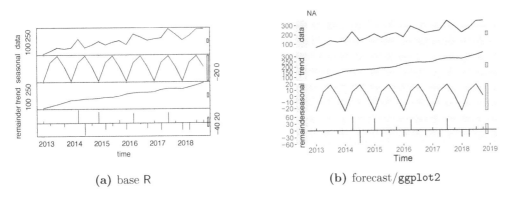

(a) base R   (b) forecast/`ggplot2`

Figure 7.10: Time series decomposition.

How to estimate the constitute components of a time series? R provides the `stl()` function, for Seasonality and Trend with Loess, an non-linear estimation procedure. A call to `stl()` with the time series object as the first argument calculates the trend, seasonal and error components. Use the base R `plot()`, or use `ggplot2` via `autoplot()`, to plot the time series of the components.

loess() function base R, Figure 5.4, p. 107

R Input *Decomposition of a seasonal time series*

data: see previous example

```
decomp <- stl(Sales.ts, s.window="periodic")
```

> *base R*: plot(decomp)
>
> *forecast*: autoplot(decomp)

Save the output to an R object to display at the console the exact numerical values of the components by entering the name of the object. Indicate seasonality is present by setting the s.window parameter to "periodic". Here the saved object *decomp* contains the numerical information from which to construct the visualization. Listing 7.9 shows only the information for the first four quarters. For the remaining years the seasonal components stay the same for each quarter, but the trend continually increases.

```
 1  > decomp
 2   Call:
 3    stl(x = Sales.ts, s.window = "periodic")
 4
 5   Components
 6                 seasonal       trend    remainder
 7   2013 Q1  -24.7730012   79.34620    9.2468021
 8   2013 Q2    7.0882904   96.49518   -9.0634701
 9   2013 Q3   17.9510934  114.78091    1.3579997
10   2013 Q4   -0.2660457  135.44925  -14.6432073
11   ...
```

Listing 7.9: Numerical output of stl().

The observed data values can be expressed in terms of this decomposition. For example, consider the first reported sales for Q1, 2013, the first data value in the sequence, 63.82. From Listing 7.9, the contribution of Trend for the Q1 of 2013 is 79.356, but its effect reduced 24.773 by negative additive component of seasonality.

$$Y_1 = T_1 + S_1 + e_1 \tag{7.1}$$
$$63.82 = 79.356 - 24.773 + 9.246 \tag{7.2}$$

This reconstitution of the data value from its estimated additive components provides the basis for the forecast into future years based on the increasing trend and the seasonality.

7.3.3   Generate a Forecast

standard error bands,
Figure 5.4, p. 107

The forecast package (Hyndman & Khandakar, 2008) provides elegant and straightforward functions by which to generate forecasts from time series data. Here two widely used forecasting methods are illustrated, the time series decomposition from the previous section as well the Holt-Winters seasonal exponential smoothing method, explained in many references including the online textbook that accompanies the forecast package (Hyndman & Athanasopoulos, 2018). The forecast functions also easily provide for a visualization of the forecasts, including standard error bands, 0.95 and 0.80 by default.

The results for the STL and Holt-Winters forecasts appear in Figure 7.11.
To create the forecasts, begin again with an R time series object, created with ts(). Refer to the corresponding forecast function, here stlf() or hw(), which

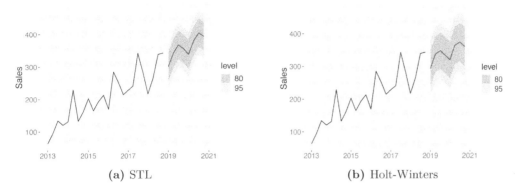

Figure 7.11: Time series forecasts.

rely upon the respective base R functions `stl()` and `HoltWinters()`. Then specify within the function calls the parameter `h`, the number of periods for forecasting. To forecast two years out, set `h` at 8 as each year has four quarters. Then save the result, here *fit1* and *fit2*, respectively, to plot, and also to inspect the specific numerical estimates upon which the corresponding visualization is based. Plot with `autoplot()`.

R Input *Time series forecast*

```
data: d <- Read("http://lessRstats.com/data/Sales.xlsx")
      Sales.ts <- ts(d$Sales, start=c(2013,1), frequency=4)

stfl: fit1 <- stlf(Sales.ts, h=8)
      autoplot(fit1, xlab=NULL, ylab="Sales", main=NULL)
hw: fit2 <- hw(Sales.ts, h=8)
      autoplot(fit2, xlab=NULL, ylab="Sales", main=NULL)
```

Enter *fit1* or *fit2* at the console to view the numerical estimates, not only for the forecast, but also for the 80% and 95% confidence interval for each quarter. Figure 7.12 displays the forecasted values themselves. The data ends in the fourth quarter of 2018, so a two year forecast begins in Q1 of 2019 through Q4 of 2020.

Forecast	2019 Q1	2019 Q2	2019 Q3	2019 Q4	2020 Q1	2020 Q2	2020 Q3	2020 Q4
STL	301.07	341.98	367.73	356.72	339.07	379.98	405.72	394.71
HW	293.23	337.03	347.37	333.59	319.62	363.41	373.76	359.98

Figure 7.12: Two year forecast.

As can be seen from Figure 7.12, the forecasted values from the two forecasting methods are similar, but not identical.

ggplot stacked time series plot,
Section 7.2.2, p. 165

Chapter 8

Visualize Maps and Networks

8.1   Maps

With the R system, create maps of the world, countries, counties, cities, or smaller areas. Add additional information to a map such as population density or one of a variety of economic indicators.

8.1.1   Map the World

Projections

projection:
Transformation to represent spherical coordinates on a flat surface.

The Earth is almost a sphere, more accurately, but still not precisely, an ellipsoid. Yet we typically view our maps on flat surfaces such as paper or computer screen. A *projection* transforms spherical coordinates, which describe specific locations on a spherical-type surface, to the x,y (Cartesian) coordinates of a flat surface. A projection flattens the coordinates of latitude and longitude of the Earth's three-dimensional surface into two-dimensions.

All maps of any part of the Earth are projections. Unfortunately, a flat surface cannot represent a spherical object without distortion. Any such transformation cannot simultaneously retain accuracy of area, direction, distance, and shape. Each possible projection compromises at least one of these properties.

Figure 8.1 presents two world maps, derived from two different projections.

(a) Lat/long Projection   **(b)** Mollweide Projection

Figure 8.1: World maps created from the same data but two different projections.

Figure 8.1a presents one of the most straightforward projections. Here map latitude and longitude are plotted directly into coordinates in the x-y plane, a version of what is called the Plate Carrée projection, which means "square plane" in French. The Plate Carrée projection renders the Earth as a flat rectangle, necessarily resulting in much distortion.

The Mollweide projection in Figure 8.1b renders a more sophisticated map of the Earth with much less distortion of the sizes of the continents. There is, however, some loss of accuracy of angles between the continents and their shapes. The Mollweide projection follows from a non-linear transformation of the spherical coordinates of latitude and longitude into the coordinates of the map's x-y plane.

Many projections are available for online viewing.

Interactive world maps from different projections
_____

URL: `https://www.jasondavies.com/maps/transition/`
URL: `http://projections.mgis.psu.edu`
URL: `http://flexprojector.com`

Or, explore further and create custom projections with the free cross-platform downloadable Java software from `flexprojector.com`.

Spatial data

To construct a map, download the corresponding *spatial data*, which describe geophysical features and locations. Examples include the boundaries between countries, altitudes, and shorelines. Store spatial data in a file called a *shapefile*. A comprehensive, free, public domain spatial dataset, provided in three different levels of detail and frequently updated, is Natural Earth, found on the web.

spatial data: Data that identify the geographic location of features and boundaries on Earth.

> `http://www.naturalearthdata.com`

Directly access this spatial data with the R package `rnaturalearthdata`. The `rnaturalearth` `ne_download()` function downloads spatial data directly from the Natural Earth website into an R data frame. Or, access data already downloaded with `ne_country()`, or, for more detail, `ne_states()`. The default value of the parameter `scale`, `"small"`, provides the least detail, but also the fastest computations.

shapefile: Vector data storage format that stores the location, shape, and attributes of geographic features.

Figure 8.1 results from setting `scale` to `"medium"`. The most detail setting requires much more processing time, set with `"large"`, which also requires installing the `rnaturalearthhires` package to access the more comprehensive data set.

Two `rnaturalearth` package functions for accessing Natural Earth spatial data
_____

rnaturalearth: `world <- ne_countries(scale="medium", returnclass="sf")`
rnaturalearth: `world <- ne_download(scale="medium", type="countries",`
                                    `returnclass="sf")`

This example obtains spatial data for the entire world because no specific countries or continents were specified.

The spatial data frame in this example, *world*, organizes the data by one of two shapefile formats: `"sp"` or `"sf"`. The `"sp"` format, for SpatialPolygonsDataFrame, is the older standard, of which geographers are transitioning away from to the newer, simpler standard, `"sf"`, or "simple features". The features are the objects described by a geometry such as points, lines, or polygons, according to vector (drawing) attributes. Each row of `sf` data represents a single spatial object such as a line, any associated data such as length, and a variable that contains the coordinates of the object.

simple features data frame example, Figure 8.4, p. 190

The functions `ne_download`, `ne_countries()` and `ne_states()` return a default `sp` spatial data frame. To obtain the emerging standard, invoke the `returnclass` parameter set to "sf". Many plotting functions, such as provided by ggplot2, convert the spatial data to "sf" format automatically, but more straightforward and faster to begin with the analysis with a "sf" data structure.

Create the world maps

For a ggplot2 map, visualizes simple feature objects with geom_sf(). Given this *geom* and the *world* data set, creating the default world map is straightforward.

Create the default version of Figure 8.1a.

---

ggplot2: ggplot() + geom_sf(data=world)

---

cartographer:
Specialist who
prepares maps from
geographic data.

The default projection in Figure 8.1a is not one that a cartographer would typically choose to represent the Earth. The Mollweide projection presented in Figure 8.1b more accurately depicts the Earth's surface, with many more alternative, sophisticated projections from which to choose. How to realize these options? Via ggplot2 the answer is the coord_sf() function, which includes the crs parameter for Coordinate Reference System (CRS). A CRS transforms geospatial coordinates from one coordinate reference system to another, which includes projections.

The following code generates both Figures 8.1a and 8.1b from the previously created *world* data frame. To provide for the flexibility of creating maps with different projections, here save the output of ggplot2 function calls to the ggplot2 object *p*. Subsequently, reference *p* to add a specific projection for each map.

R Input *Create the world maps in Figure 8.1*

```
ggplot2: p <- ggplot() + geom_sf(data=world) +
         theme_set(theme_bw()) +
         theme(panel.grid.major = element_line(color="gray75", size=.5))
Lat/long: p + coord_sf(crs="+proj=longlat")
Mollweide: p + coord_sf(crs="+proj=moll")
```

To map only part of the world use parameters xlim and ylim for function coord_sf() to specify a vector of starting and ending longitudes and latitudes, respectively.

The CRS transformations are from the public domain PROJ library (PROJ contributors, 2018), version 4, written as PROJ.4. The sf function st_proj_info() lists the available projections within R. The following website references some of the official documentation of PROJ, visually displays each projection, and describes its parameters.

Available projections in R

---

sf: st_proj_info(type="proj")
URL: https://proj4.org/operations/projections/index.html

---

proj-string:
Description of a
coordinate system
renders a map.

Within the PROJ system, describe coordinate transformations with *proj-strings*, which serve as values of the crs parameter for the ggplot2 function coord_sf(). Precede each parameter in a proj-string with a `+`. Focus on the projection parameter, `proj`. The default value is `longlat`, explicitly provided in the preceding `ggplot2` function call. The value of `proj` for the Mollweide projection is `moll`.

Also required is a model, called a *datum*, of the specific spherical surface that serves as a reference point for the geospherical coordinates longitude and latitude. The most common choice of cartographers is the *WGS84* datum, the 1984 World Geodetic System standard, which also serves as the reference standard for GPS, the Global Positioning System. Still, alternate Earth surface models are available, as well as models designed specifically for local regions.

> **datum**: Define the shape and size of the Earth and provide a reference point for describing locations via coordinates.

The reliance upon *WGS84* can be made explicit with `st_crs()` from the `sf` package, which retrieves the coordinate reference system upon which an `sf` data frame is based, illustrated for the *world* data frame in Listing 8.1.

```
> st_crs(world)
Coordinate Reference System:
  EPSG: 4326
  proj4string: "+proj=longlat +datum=WGS84 +no_defs"
```

Listing 8.1: Apply the `st_crs()` function to the *world* `sf` data frame.

The EPSG number is an alternate designation of the obtained prog-string. Each spatial data set is only interpreted within the context of a datum, and the projection that defines the map that displays the data. Choose different datums to enhance the accuracy of a specific location. Using the correct datum on which the spatial data is based is essential because distances between locations can substantially differ according to different datums.

8.1.2 Raster Images

Store an image in one of two primary formats: vector or raster. A *raster image* consists of many pixels (on a computer screen) or dots (on a printed image). In contrast, a *vector image* consists of a mathematical description of geometric objects such as points, lines and polygons, and their relative positioning and sizes. A photograph is a raster image, typically composed of a large number of tiny pixels or dots. The format can require much storage space and does not scale well to larger images, but can achieve much detail and gradation of colors. Vector images can scale perfectly, but without the fine level of detail and gradation provided by a raster image.

> **raster image**: Set of tiny pixels or dots, each of which conveys a color.
>
> **vector image**: Set of mathematical descriptions of geometric objects and positioning.

The previous examples of spatial data files in `sf` format are vector files. The `ne_download()` function also provides for highly detailed raster images available as part of the Natural Earth data set, such as Figure 8.2.

To create this raster map, first download the data with the previously introduced `ne_download()`, but with the `category` parameter set to `"raster"`. Set the `type` parameter to `"MSR_50M"` to indicate a medium scale raster Manual Scale Relief map. Here create the map with the base R function `plot()`, applied to objects of the class `raster`. As such, the full name of the plotting function is `plot.raster()`,

Figure 8.2: Grayscale raster image scale relief map of the world from Natural Earth.

invoked with `plot()`.

R Input

naturalearth: r.world <- ne_download(scale="medium", type="MSR_50M",
                                             category="raster")

base R: plot(r.world, col=getColors("black", "white", n=48)

Create the grayscale image with the parameter `col`. In this example, the `lessR`
function `getColors()` generates a sequential scale of 48 shades of gray from `"black"`
to `"white"`. The default color palette for `plot.raster` generates the base R terrain
colors. The following call to `getColors()` displays the terrain palette.

getColors() `lessR`,
Section 10.2.1, p. 215

Base R terrain palette

lessR: getColors("terrain", n=100, border="off")

Of course, create the raster image map with any valid color sequential scale.

8.1.3   Online Geocode Databases

Geocoding provides the geographical coordinates, latitude and longitude, of a specific
location, its *geocode*. Locate any place on the Earth's surface, then plot the
corresponding location on a map.

geocoding: Obtain
the geographical
coordinates for a
given location.

geocode: The
geographical
coordinates of a
location.

US Census Bureau geocodes

Multiple commercial options provide geocoding, but so do some quality free alternatives. For USA locations, the US Census Bureau offers a free, authoritative source
of geocodes, an extensive database of addresses and their corresponding longitudes
and latitudes.

US Census Bureau geocodes website

URL: https://geocoding.geo.census.gov

To obtain an individual geocode enter an address at the prompt. For the One Line
button enter as a single line with comma delimiters. For the Address button, enter
multiple lines. Or, click on the Address Batch button to submit a batch file of up

to 10,000 lines of addresses in either `csv` or Excel format. The geocodes are then returned as a file in the same format as submitted. The first column of the input file is the row number. The `csv` input file cannot end with a blank line.

Google Map Services

Google Map Services provides for international geocodes, with free limited use. Although the lookup of geocodes costs .005¢ per geocode, or $5.00 per 1000, a $200 credit automatically applies each month. Without the use of other mapping services, the first 40,000 geocodes per month are functionally free. The downside is that, unlike the previously discussed geocoding source, obtaining a geocode costs money, even if refundable, and so can only be accessed by providing Google a credit card number.

To access this service within R requires registration at Google map services, and then proof of such registration. Register at the `cloud.google.com` website to obtain an Application Programming Interface (API) key to access specific types of information, here the geocode API as well as five other mapping API's by default. The second listed website is the source for maintaining the account.

> *Google map service geocodes registration*
>
> *URL*: `https://cloud.google.com/maps-platform/`
> *URL*: `https://console.cloud.google.com`

To register at the website, click a Get Started button, then choose Places. Next, select an existing project or enter a new project name. Enter credit card information for billing. The provided long character string called a personal API Key allows access to the mapping service from within R.

To inform R of your personal key, load the `ggmap` package with `library()`, and then `register_google()` with the `key` parameter. The related `has_google_key()` function indicates if a key is successfully registered, available for the current R session.

library() base R, Section 1.1.2, p. 6

> *Inform R of your personal key*
>
> *ggmap*: `register_google(key="personal_key")`
> *ggmap*: `has_google_key()`

The `register_google()` function parameter `write` is `FALSE` by default. If set to `TRUE`, the function writes the entered personal key into the R file system, automatically available for future use. Unfortunately, R does not store confidential information, so that it may be possible for a rogue package to locate this key and transmit the information somewhere else. Alternatively, obtain the personal key from the google console, from the `URL` provided above.

Given a registered API key, Listing 8.2 illustrates the use of `geocode()` from `ggmap` to build a data frame of three locations with longitudes and latitudes. First, create a character vector of the locations. Then `geocode()` submits the locations to the

vector character, Section 1.2.4, p. 11

mapping service, which returns the latitudes and longitudes. Merge the data frame of locations with their corresponding coordinates, such as with the base R "column bind" `cbind()` function.

```
location <- c(
"615 SW Harrison St, Portland, OR",
"Disneyland",
"Rio de Janerio, Brazil"
)
d <- geocode(location)
d <- cbind(location, d)
d
                          location      lon      lat
1 615 SW Harrison St, Portland, OR -122.6832  45.51156
2                        Disneyland -117.9190  33.81209
3            Rio de Janerio, Brazil  -43.1729 -22.90685
```

Listing 8.2: Input, function call, and output for package **ggmap** function `geocode()`.

The example in Listing 8.2 demonstrates `geocode`'s impressive flexibility to identify locations according to a variety of references. The first location is a standard address, the second the name of a well-known landmark, and the third only a city name.

Geonames geocodes

A free, public domain database, GeoNames, offers city geocodes that include cities across the world. The database includes cities with as few as 500 or more inhabitants, as well as 1000 or more, 5000 or more, and 15000 or more inhabitants.

Readme file for the Geonames databases

URL: http://download.geonames.org/export/dump/readme.txt

Licensed under the generous Creative Commons Attribution 4.0 License, the data and related information can be copied and redistributed in any medium or format, and transformed and modified as needed for any purpose, even commercially, with the requirement to cite the source of the data.

Download the zipped data file from *geonames.org* with base R `download.file()`. Provide the URL and the destination file. Because no path name precedes the file name in this example, R writes the file to the current working directory, obtained from `getwd()`. Base R `unzip()` unzips the file to a `.txt` file to read into R. The downloaded data file, however, does not include the variable names. Specify the variable names with `Read()` as the value of the base R parameter `col.names`.

Download and read Geonames databases such as cities15000.zip

base R: download.file("http://download.geonames.org/export/dump/
                    cities15000.zip", "cities15000.zip"))

base R: unzip("cities1500.zip")

lessR: d <- Read("cities15000.txt", col.names = c("id","name",
        "ascii_name","alt_names","latitude","longitude","feature_class",
        "feature","country.code","cc2","admin1","admin2","admin3",
        "admin4","population","elevation","dem","timezone","mod_date"))

Once read, query the data frame for the specific city geocodes of interest. For example, apply base R subsetting to limit the *d* data frame to all Italian cities larger than 250,000 inhabitants given the specified variables.

R Input *Base R query of geocities database for specified cities, variables*
data: from above

```
cols <- c("name", "longitude", "latitude", "population", "elevation")
rows <- d$country.code=="IT" & d$population > 250000
d <- d[rows, cols]
d
```

Listing 8.3 reports the query results.

```
          name longitude latitude population elevation
11990   Palermo  13.33561 38.13205    648260        14
12020   Catania  15.07041 37.49223    290927         7
12084     Turin   7.68682 45.07049    870456       239
12162      Rome  12.51133 41.89193   2318895        20
12231    Naples  14.26811 40.85216    959470        17
12258     Milan   9.18951 45.46427   1236837       122
12330     Genoa   8.94439 44.40478    580223        19
12354  Florence  11.24626 43.77925    349296        50
12458   Bologna  11.33875 44.49381    366133        54
12470      Bari  16.86982 41.12066    277387         5
```

Listing 8.3: Extracted geocode information from free, downloadable Geonames database.

Define one vector, *cols*, to specify the variables to retain, and another vector, *rows*, to specify the rows of the data frame to retain. Then subset the data, such as with the base R Extract operator, [ ]. Replace the *country.code* in the definition of *rows* to select cities from any other country.

subsetting (extract) function, Section 1.2.4, p. 13

8.1.4 Create a Country Map with Cities

Here create a map of Italy with its ten most populous cities, Figure 8.3. The corresponding `ggplot2` code references different data sets in different layers. One data set defines the polygons that compose the map of Italy. The second data set provides the geographical coordinates and populations for the ten cities.

Create the data frame of polygons for Italy and its provinces with the `rnaturalearth` function `ne_states()`.

The `ne_states()` function requires package `naturalearthhires`. Obtain with:

```
install.packages("devtools")
install_github("ropensci/rnaturalearthhires")
```

create data frame of Italian cities, Listing 8.3, p. 187

Set `country` to `"italy"`. Obtain the city coordinates from the already obtained *d* data frame of the ten most populous Italian cities and their geographical coordinates.

R Input *Two data sets for the map of Italy given data frame d from Listing 8.3*
rnaturalearth: `italy <- ne_states(country="italy", returnclass="sf")`

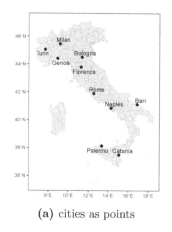

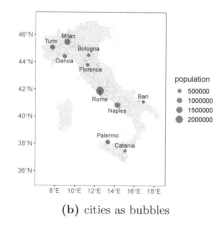

(a) cities as points   **(b)** cities as bubbles

Figure 8.3: Italy and its ten most populous cities.

```
sf:   cities <- st_as_sf(d, coords = c("longitude", "latitude"),
                         crs=st_crs(italy), remove=FALSE)
```

Multiple spatial data sets plotted on the same panel should all share the same coordinate reference system and projection. If not explicitly specified with `coord_sf()`, ggplot2's `geom_sf()` assumes the coordinate reference system of the first set, then transforms as needed on subsequent layers to ensure a common reference system. `geom_sf()` also adds the labeled lines of longitude and latitude.

However, here use `st_as_sf()` from the **sf** package to assign the *italy* CRS to the cities data frame *d*. For parameter `crs`, invoke `st_crs()` to retrieve this CRS. Set the `st_as_sf()` `remove` parameter to `FALSE` to retain the city latitude and longitude as separate variables in the resulting **sf** data frame.

world maps example,
Section 8.1.1, p. 182

Plot polygons for the map of Italy, and plot points for the cities placed on the map with `geom_sf()`. Constrained to grayscale, set the `fill` for the polygon interiors and `color` for the borders to shades of gray. Reduce the size of the border lines from their default by setting `size` to 0.2. Increase the default plot `size` to 2 for the size of the cities.

The `geom_text_repel()` function, from the **ggrepel** package, provides convenient features for maps. By default, `geom_text_repel()` writes the text, here the city names, above or below the corresponding point. As always, first retrieve the package from the R package library with `library()`.

library() base R,
Section 1.1.2, p. 6

R Input *Map of Italy with its ten most populous cities plotted as points*
```
ggplot2: ggplot() +
     geom_sf(data=italy, fill="gray85", color="gray65", size=0.2) +
     geom_sf(data=cities, size=2) +
     theme_set(theme_bw()) + labs(x=NULL, y=NULL) +
ggrepel: geom_text_repel(data=cities, aes(longitude,latitude, label=name),
                         size=3.25, col="black")
```

Figure 8.3b plots the coordinate for each city as a bubble, the size of which depends on the population of the corresponding city.[1] The `scale_size_area()` function scales the population as the area of each bubble, instead of the default radius. Activate plotting the bubble plot by mapping the variable *population* to the parameter `size`.

scale_size_area() `ggplot2`, Section 5.2.3, p. 118

The `geom_text_repel()` parameters `nudge_y` and `nudge_x` move the corresponding text element the specified amount. If the text element becomes too far from the corresponding coordinate, `geom_text_repel()` automatically draws a line segment from the text to the coordinate. Apply the `nudge` parameters to an array of the same length as the number of text elements to plot. Each element corresponds to the corresponding coordinate in the defining data frame. Here also invoke the `size` parameter to increase the text from its default size.

> **R Input** *Map of Italy with its ten most populous cities plotted as bubbles*
>
> ```
> ggplot2: ggplot() +
>       geom_sf(data=italy, fill="gray85", color="gray65", size=.2) +
>       geom_sf(aes(size=population), data=cities, alpha=.7) +
>       scale_size_area() +
>       theme_set(theme_bw()) + labs(x=NULL, y=NULL) +
> ggrepel: geom_text_repel(data=cities, aes(longitude,latitude, label=name),
>                     size=3.75, col="black",
>                     nudge_y=c(.5,.4,.5,-.6,-.4,.4,-.2,0,.5,.4))
> ```

Visualize any other country on Earth, or multiple countries by specifying a vector of countries, with the `rnaturalearth` function `ne_states`.

8.1.5   Choropleth Map

A *choropleth map* shades areas according to the values of a variable. One such variable is the Gini coefficient (Gini, 1921), an index of income inequality. Figure 8.4 illustrates a choropleth map for the 48 contiguous USA states of 2017 Gini coefficient data from the US Census Bureau (2017). From the map, New York state has the largest income equality, followed by a band of states throughout the southeast and California. The northwestern and midwestern states have the lowest income inequality.

choropleth map: Shade or color specific areas colored or shaded in proportion to the value of a corresponding variable.

This map relies upon another widely used source of mapping data, the `maps` package. The map data files produced by the `map` function from the `maps` package are not formatted as simple features, so convert with `st_as_sf()`. Store the result in the `sf` data frame *states*. Listing 8.4 presents the first several lines of *states*. After the header information, the variable *geometry* in the converted data frame encodes the information for the polygon that defines the boundary of each state. The variable *ID* encodes the state names, all in lower case.

Find the *gini* data at the American Community Survey data section of the US Census website, available at `factfinder.census.gov`. Remove the first row and a

---

[1]The `geom_sf()` function properly plotted the bubbles, but due to an apparent bug, did not properly display the legend. As such, `geom_point()` did the plot in Figure 8.3b according to: `aes(longitude, latitude, size=population)`.

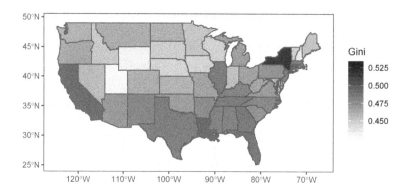

Figure 8.4: Income inequality by State of USA, 2017 Gini coefficients.

```
> head(states, n=3)
Simple feature collection with 3 features and 1 field
geometry type:  MULTIPOLYGON
dimension:      XY
bbox:           xmin: -114.8093 ymin: 30.24071 xmax: -84.90089 ymax: 37.00161
epsg (SRID):    4326
proj4string:    +proj=longlat +datum=WGS84 +no_defs
                          geometry      ID
1 MULTIPOLYGON (((-87.46201 3...  alabama
2 MULTIPOLYGON (((-114.6374 3...  arizona
3 MULTIPOLYGON (((-94.05103 3... arkansas
```

Listing 8.4: Header information and first three rows of data for the simple features *states* data frame.

few columns of the downloaded file not relevant to the current analysis. To match the information in the *states* data frame, covert the state names to lower case and rename the column of State names as *ID*. Rename the column of Gini scores as *Gini*.

To create the Figure 8.4 map, merge the Gini data, which exists in a regular data frame, into the special features *states* data frame. The *states* data frame consists of 49 rows of data, the contiguous 48 states and the District of Columbia. The *gini* data frame consists of all 50 states plus the District of Columbia and Puerto Rico.

inner join: Merge two data structures by retaining only rows of data with shared values of a join variable (ID) in both structures.

Merge the data frames with an inner join according to shared values of *ID*. Use the **sf** package function **inner_join.sf()**, which follows the form of the tidyverse **dplyr** function, **inner_join()**. No need to include the reference **sf**. The result is a simple features data frame with the Gini coefficient for each state that exists in both of the pre-merged data frames.

R Input *Data wrangling to create the data frame for analysis yielding Figure 8.4*

sf, maps: `states <- st_as_sf(map("state", plot=FALSE, fill=TRUE))`
lessR: `gini <- Read("http://lessRstats.com/data/Gini2017.xlsx")`
base R: `gini$ID <- tolower(gini$ID)`
sf: `states <- inner_join(states, gini, by="ID")`

To create the map, specify the *states* data frame as the input data to `geom_sf()`. Then map the Gini coefficient to the visual aesthetic of choice, here parameter `fill`. The optional second line of code, the call to `scale_fill_gradient()`, displays the map in grayscale. The default is a sequential blue palette.

> **R Input** *Mapping code to create Figure 8.4*
> *ggplot2*: `ggplot() + geom_sf(data=states, aes(fill=Gini)) +`
> `    scale_fill_gradient(low="gray95", high="gray5")`

Visualizing the distribution of a continuous variable, such as the Gini coefficient, provides information complementary to that provided by a map. How is the Gini coefficient distributed across the states?

Consider the VBS plot (Gerbing, 2020) from `lessR Plot()`, the integrated violin plot, box plot, and scatterplot. Figure 8.5 reveals an approximately normal distribution of Gini. A second Gini coefficient, for New York state, approaches outlier status as it is near but not beyond the fence, the boundary for labeling a point as an outlier.

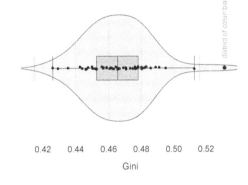

Figure 8.5: VBS plot of the Gini coefficient of income inequality across the 48 contiguous states of the USA plus the District of Columbia.

VBS plot,
Section 4.5, p. 94

8.2 Network Visualizations

A *network* consists of nodes and edges that join the nodes. A social network represents people as nodes. The edges represent relationships among the people. An information network consists of information flows between people or organizations, such as a network of employee email communication. A biological network of feeding relationships from plants to predators, a food web, shows which organisms feed off of others. Transportation networks indicate routes that connect locations.

network: Set of objects, nodes, joined by edges that represent relationships.

A network visualization reveals the interconnected nodes of the network. What nodes are central? What sub-groups exist? For a transportation network, what is the shortest path through the network?

R provides several traditional quality network visualization tools that include the `igraph` (Csardi & Nepusz, 2006) and `network` (Butts, 2008) packages. A traditional representation of a network, used by both `igraph` and `network`, is an *adjacency matrix*, a square matrix with the columns and rows named according to the nodes. Each data value of 1 within the matrix indicates that there is a direct connection between the respective row and column nodes. A 0 indicates no connection.

adjacency matrix: Representation of network connections with a square matrix of nodes with data values of 1 or 0 for a connection or not.

A more recent development represents network data in terms of data frames or tibbles so that traditional data manipulation tools can be applied to network data, such as Thomas Lin Pedersen's `tidygraph` (2019) package. Pedersen's complementary `ggraph` (2018) package works with syntax similar to `ggplot2` to create the network graphs from the **tidygraph** data structure. The visualization software also accesses more traditional network structures, such as from **igraph**, by converting them to the tidygraph structure.

8.2.1 Network Data

asymmetric relation: The direction from one node to another does not exist in the return direction.

Consider four cities, in Figure 8.6 generically named City_A through City_D. City_C is central, and directly connects with the other three cities. The network represents commuter trains that travel into the city for a morning commute, in which some of the routes are *assymetric* or directed. Bi-directional service at that time is available only for Cities A and D and Cities B and C.

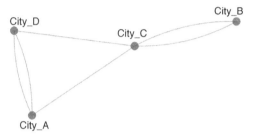

Figure 8.6: Demonstration network.

<u>Data method 1</u>. The **ggraph** network visualization data structure is a **tbl_graph**, a composite tibble data frame of two component tibbles: one for the nodes and one for the edges. The node data frame contains a column to identify each node, *name* in this example. The edge data frame requires *from* and *to* variables to identify the nodes that each edge connects.

Both node and edge data structures may also contain other variables. In this example, the cities reside in two different transportation zones. The zone number is recorded with the node names as the variable *zone*. The average duration of each trip along each route is recorded in the edges data frame as the variable *time*.

Listing 8.5 reveals the structure of the four-node network expressed as nodes and edges data frames.

```
> the_nodes              > the_edges
    name zone                 from       to time
1 City_A    2            1 City_A City_C   22
2 City_B    1            2 City_B City_C   16
3 City_C    1            3 City_D City_C   24
4 City_D    2            4 City_A City_D   44
                         5 City_C City_B   13
                         6 City_D City_A   41
```

Listing 8.5: Nodes and edges data frames.

Find these two data frames from Listing 8.5 on the web in a single Excel file with two worksheets. The first sheet defines the nodes. The second sheet defines the edges. After reading the data into their respective data frames, combine them to form the composite **tbl_graph** data structure with the **tidygraph** function **tbl_graph()**.

R Input *Create ggraph data structure tbl_graph from nodes and edges data frames*
lessR: the_nodes <- Read("http://lessRstats.com/data/gr_NodesEdges.xlsx")

```
    the_edges <- Read("http://lessRstats.com/data/gr_NodesEdges.xlsx",
                      sheet=2)
```

tidygraph: `network <- tbl_graph(nodes=the_nodes, edges=the_edges,`
`                       directed=TRUE)`

Listing 8.6 shows the structure of the composite tibble data frame, the `tbl_graph` named *network* in this example. The information contained in the separate nodes and edges data frames combines into a single tibble data frame.

```
> network
# A tbl_graph: 4 nodes and 6 edges
#
# A directed simple graph with 1 component
#
# Node Data: 4 x 2 (active)
  name    zone
  <chr>   <int>
1 City_A     2
2 City_B     1
3 City_C     1
4 City_D     2
#
# Edge Data: 6 x 3
   from     to  time
  <int> <int> <int>
1     1     3    22
2     2     3    16
3     4     3    24
# ... with 3 more rows
```

Listing 8.6: Nodes and edges data frame of type `tbl_graph`.

Data method 2. The alternate approach to forming the `tbl_graph` data structure begins with a data structure that contains all the needed node and edge information, but not in the `tidygraph` format. The `tidygraph` function `as_tbl_graph()` converts the information into the nodes and edges tibble data frames. Then combine into a composite tibble data frame with `tbl_graph()`.

The function `as_tbl_graph()` converts many different data structures, which include a standard data frame that contains the edges from which `as_tbl_graph()` can deduce the `tbl_graph()` structure. Manipulate this structure to add additional variables that describe the nodes, such as the zones in this example. Other examples include structures from packages such as `igraph`, `network`, or other alternatives that include the output of R's hierarchical clustering function `hclust()`.

hclust() hierarchical clustering base R function, Figure 5.18, p. 124

From the `tbl_graph` data structure, here named *network*, visualize with `ggraph()`.

8.2.2 Visualizations

The following figures present different visualizations of this simple four-city network. The sparse Figure 8.7a is the default `ggraph()` visualization. Figure 8.7b builds upon the default with labeled nodes. `ggraph()` randomly orients each created visualization, so these two visualizations orient differently.

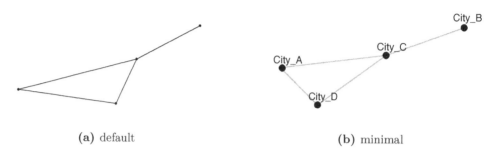

(a) default   (b) minimal

Figure 8.7: Default and minimal `ggraph` network visualizations.

To create Figure 8.7a, call `ggraph()` with the network data frame object of type `tbl_graph`, here named *network*. The geom `geom_edge_link()` plots the edges and `geom_node_point()` plots the nodes. List the node `geom` after the edge `geom` to display the nodes over the edges. The `theme_graph()` call removes the gray default `ggplot2` background and generally yields a more appropriate network visualization.

> **R Input** *Create Figure 8.7a default network diagram*
>
> ```
> ggraph(network) + geom_edge_link() + geom_node_point() + theme_graph()
> ```

Adding some parameter values results in the more useful Figure 8.7b visualization. For `geom_edge_link()`, set `width` to 1 to widen the edges, and set `color` to `"gray70"` to lighten the color, or choose any desired color. Increase the size of each node with the `geom_node_point()` `size` parameter. The `geom_node_text()` function labels the nodes. Indicate the source of the `label` for each node from the variable `name`. The `repl` parameter performs the same role as for the `geom_text_repel()` function, from the `ggrepel` package, to position the labels. The `vjust` parameter shifts the labels vertically.

repel, Section 8.1.4, p. 188

> **R Input** *Create Figure 8.7b network diagram with larger, labeled nodes*
>
> ```
> ggraph(network) + geom_edge_link(width=1, color="gray70") +
>     geom_node_point(size=4) +
>     geom_node_text(aes(label=name), repel=TRUE, vjust=-.75) +
>     theme_graph()
> ```

Figure 8.8 presents more sophisticated visualizations of this simple network. Additional features include arrows to indicate directionality, and the width of the edges reflecting the amount of time required to traverse the paths. Figure 8.8b includes those times as labels on the edges as well. Nor are the edges straight lines.

Create more sophisticated edges with the `ggraph` family of edge functions. The straight edges in Figure 8.7 result from `geom_edge_link()`. The Figure 8.8 visualizations rely upon `geom_edge_diagonal()` and `geom_edge_fan()`, respectively. The term "diagonal" does not refer to diagonals as in a rectangle, but to a type of wavy bezier curve. The `geom_edge_fan()` generalizes to asymmetric relationships with edges in both directions.

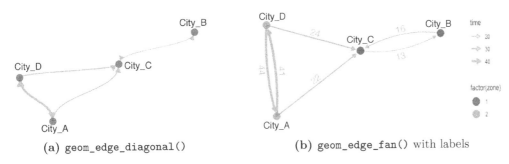

(a) `geom_edge_diagonal()`   (b) `geom_edge_fan()` with labels

Figure 8.8: Network `ggraph` visualizations with alternative edge geoms that include asymmetric relationships.

The input `tbl_graph` data structure, *network*, contains a node variable *time*, the number of minutes required to travel from one node to an adjacent node. Map *time* to the `width` parameter in the edge function call to `aes`. Specify arrowheads with the `arrow` parameter, which calls the `arrow()` function, here with a 3mm closed arrowhead. The `end_cap` parameter specifies to end the arrowhead 2mm before the center of each node, instead of over the node. The provided start and end values of the parameter `range` for the `scale_edge_width()` function scales the width of the edges according to the value of `width`.

R Input *Create Figure 8.8a with curved edges, arrowheads, and custom edge widths*

```
ggraph(network, layout="nicely") +
  geom_edge_diagonal(aes(width=time), color="gray65",)
    arrow=arrow(length=unit(3,"mm"), type="closed"),
      end_cap=circle(2,"mm")) +
  scale_edge_width(range=c(0.4, 1.75)) +
  geom_node_point(color="gray30", size=5, alpha=.8) +
  geom_node_text(aes(label=name), repel=TRUE) +
  theme_graph()
```

Figure 8.8b also displays the label for each edge, here transit time. Map the variable *time* to the `label` visual aesthetic in the call to `geom_edge_fan()`. Also, map the `factor` version of *zone* to the `color` visual aesthetic, scaled so that Zone 1 cities have darker gray nodes than Zone 2 cities. The third mapping maps the *name* of each node to the `label` aesthetic.

map data values,
Section 10.2.1, p. 216

Choose from several different network algorithms to calculate the placement of the nodes. The `ggraph` code underlying Figure 8.8 defines the algorithm that lays out the visualization with the `layout` parameter, here set to the default `"nicely"`. Many such algorithms exist, offered directly with the `ggraph` package, or indirectly via the `igraph` package. Listing the name of one of the many `igraph` layouts as the argument to `layout` invokes the specified layout. Enter `?layout_igraph_auto` into the R console to display the available `igraph` layouts.

By default, `ggraph()` positions the edge labels centered over the edges. To place
along the edges, set parameter `angle_calc` to `"along"`. To nudge the labels up a
little beyond the edges, set `label_dodge` to 3mm.[2]

> **R Input** *Create Figure 8.8b with two-way, labeled edges, and custom edge widths*
> ```
> ggraph(network, layout="nicely") +
>     geom_edge_fan(aes(width=time, label=time), color="gray65",
>       arrow=arrow(length=unit(2,"mm"), type="closed"),
>       end_cap=circle(3,"mm"), angle_calc="along",
>       label_colour="gray60", label_dodge=unit(3,"mm")) +
>     scale_edge_width(range=c(0.4, 1.75)) +
>     geom_node_point(aes(color=factor(zone)), size=5, alpha=.8) +
>     scale_color_manual(values=c("gray20", "gray50")) +
>     geom_node_text(aes(label=name), repel=TRUE, vjust=-2) +
>     theme_graph()
> ```

The nodes, edges, and labels can be displayed with customized colors. Use `color`
for the nodes and edges, as illustrated in the preceding code. For the labels, use
`label_colour`.

8.2.3   Network Analysis

Many tools exist for analyzing different aspects of a network, such as locating the
shortest route from one node to another, or of identifying clusters, subgroups, of
nodes. Here consider network centrality. Some nodes in a network may be more
central than others, such as a location in a transportation network embedded in
most routes. Network centrality measures provided by network software `igraph` are
also available via `ggraph`. One measure, *betweeness*, assesses the number of shortest
paths from each node to all other nodes.

betweeness of a
node: Approximately
the number of
shortest paths that
pass through the
node.

To access either the nodes or the edge data table within this composite tibble
data frame, *network*, call the `activate()` function, either `activate(nodes)` or
`activate(edges)`. Subsequent data manipulations then apply only to either the
nodes or the edges. Here apply the pipe operator to activate the node section of
network, then assess centrality with `centrality_betweenness()`, saving the result
into a created variable *central* using tidyverse `mutate()`. The `->` indicates to save
the update back into the data structure.

pipe operator,
Section 2.5.1, p. 41

> *Network centrality analysis and data manipulation*
> ```
> network %>%
>     activate(nodes) %>%
>     mutate(central = centrality_betweenness()) -> network
> ```

Listing 8.7 shows the addition of the created variable *central* to the node section
of the *network* data structure. The updated node data table now contains three

---

[2]However, for the asymmetric relations with edges in both directions, the placement algorithm
did not work, so post processing with Adobe Illustrator properly positioned the labels, and also
some minor adjustment to the node labels.

instead of two variables: *name*, *zone*, and *central*. The betweeness centrality score identifies City_C, with a *central* value of 2, as the most central node, on the shortest path from City_A to City_B as well as from City_D to City_B.

```
> network
# A tbl_graph: 4 nodes and 6 edges
#
# A directed simple graph with 1 component
#
# Node Data: 4 x 3 (active)
  name     zone    central
  <chr>   <int>      <dbl>
1 City_A      2          0
2 City_B      1          0
3 City_C      1          2
4 City_D      2          0
#
# Edge Data: 6 x 3
   from      to  time
  <int>  <int> <int>
1     1       3    22
2     2       3    16
3     4       3    24
# ... with 3 more rows
```

Listing 8.7: `tbl_graph` data structure that includes the centrality score for each node.

Entire books on network analysis provide more detail than provided here. This section introduces many core concepts of visualization and analysis applied to a simple network. These concepts straightforwardly generalize to large networks.

Chapter 9

Interactive Visualizations

9.1   Interactive Visualizations with Shiny

Interactive visualizations transform static visualizations into a portal for real-time visual discovery. Interactive R visualizations empower the analyst with no knowledge of R programming to create R visualizations. Point and click to choose variables for analysis, select sub-groups of data, or run an analysis. Users can even upload their data into a standard template to produce predefined analyses, such as a scatter plot with regression line. The R interactive output appears on web pages, viewed with the familiar web browser.

9.1.1   Static vs. Interactive Visualizations

`static visualization`: To revise, manually re-enter the name of the function, with revised variable names and parameter values, then re-run.

To create an R interactive visualization, begin with the corresponding static visualization. Choose a specific function, such as the `lessR Histogram()` or `ggplot2 geom_histogram()`. Then add interactivity such as with the R interactive procedure Shiny, directions to follow.

histogram bin width, Section 4.1.2, p. 82

For example, to provide a custom bin width with `Histogram()`, specify a value for the `bin_width` parameter. To create a histogram with a different bin width, the static visualization requires re-entering the function call to `Histogram()` with a different value of `bin_width`. For the interactive version, change the parameter values interactively. For example, Shiny provides web page interfaces such as a slider to slide the pointer to either decrease or increase a numeric value such as bin width, shown in Figure 9.1. These input values are *reactive*. Revised input automatically leads to revised output.

`Shiny`: From interactive R input visualize output within a web browser.

`reactivity`: Automatically re-compute output from changed input.

Figure 9.1: Slider to input histogram bin width.

The shape of the displayed histogram varies according to the different entered values of bin width, one after the other to explore a range of possibilities.

9.1.2   Shiny Overview

RStudio, Inc. Both Shiny and RStudio are products of RStudio, Inc., so Shiny comfortably runs within RStudio.

Develop the R code for interactive visualizations on your computer, such as with RStudio. Shiny is a client-server web application. The server runs the computations. The server computer can be your computer running RStudio, one on a local network, or a computer accessible on the Internet. To allow others access to your Shiny app, deploy on a network or web-connected server computer that runs the Shiny server software.

Shiny server `https:// rstudio.com/ products/shiny/ shiny-server/`

Upon instructions for the server, the user's computer generates a web page to which the user interacts. The user's web browser, the client, provides for the input on which the computations follow, such as the slider in Figure 9.1. The client passes the entered values to the Shiny server, which then computes the output. The server then returns the output values to the user-interface. The generated web page also serves as a placeholder for the output that results from those computations. To interact with a Shiny app, point a web browser to the URL that hosts the app.

Shiny code for a single app consists of two sections. The server code for the needed computations runs on the server computer. The user interface code, *ui*, runs on the user's computer to interact with the web page for input and the display of the resulting output. Except for small demonstration apps, named `app.R`, better to maintain the code for each section in its own file, respectively named `ui.R` and `server.R`.

1. Point a browser to the server location running the app, or if on the local computer, run the app through RStudio.

2. The user interface code, the *ui*, creates a web page with a provision to input a value, such as a bin width for a histogram computation.

3. The user-interface code via the *input* parameter passes this value to the Shiny server on whatever computer hosts the server code.

4. The `server` function processes its R code with one or more values defined in the *input* parameter.

5. The `server` function returns the output via the *output* parameter to the user-interface.

6. The user-interface then displays the newly generated output, such as the histogram with the chosen bin width, generally on the same web page that contains the input device.

To develop a Shiny app, understand how to develop the user interface code and server code, and how to relate the two, the *ui* object and the *server* function.

Shiny server

The server computer processes the R code to accomplish the analysis. As illustrated in Listing 9.1, the code for the computations is defined within a function named `server`, in Lines 6-12.

> `server`: Performs computations to generate the output, such as a visualization from the user-specified input.

```
1   library(shiny)    # access Shiny functions
2   library(lessR)    # access lessR functions
3
4   d <- Read("http://lessRstats.com/data/employee.csv")    # read data
5
6   server <- function(input, output) {    # the server function
7
8     output$myPlot <- renderPlot({    # output is plot myPlot
9       Histogram(Salary, bin_width=input$myWidth)    # my_width is input bin width
10    })
11
12  }
```

Listing 9.1: Example of `server.R` code with interactive bin width for a histogram.

The `server` function defines at least two parameters: *input* and *output*. The server function receives information from the *ui* via its *input* parameter, processes this information, and then returns the generated output to the *ui* via its *output* parameter. This two-way communication process continues while the Shiny app runs, always "listening" for new input, and then reacts to any change in the input to generate the new corresponding output.

This example is of only one interactive input value, the `bin_width` parameter passed to *input$myWidth* in Line 9, within the `server` function. The `Histogram()` function accepts the input value as its `bin_width` setting. The user interface, where the user inputs this value, also defines this same value with the same name. Because the *input* and *output* parameters are standard R lists, refer to a specific value of the list with the parameter name followed by a $ and then the value name.

This example also provides only a single *output* value, *myPlot*, *output$myPlot* by its full name, defined in Line 8. To structure the function *output* ready for input into the *ui*, present the `Histogram()` function in the body of what Shiny calls a *render function*. To structure a visualization output, invoke the `renderPlot()` function.

The general syntax of a render function generalizes from Lines 8-10 in Listing 9.1.

> **R Input** *Form of a Shiny render function*
>
> ---
>
> *output$*output_name $\Leftarrow$ **render*name*** ({ R code with *input$*input_name})

The *name* is the type of object that the R code generates. For example, `renderPlot()` defined in Lines 8-10 of Listing 9.1 specifies the R code to create a visualization. The brackets in the render function call, { and }, allow for multiple lines of R code to define the output, although here only one line defines the function. The output of this function defines at least one value of the `output` parameter.

render function: Define a value of the *output* parameter according to the type of output generated.

Table 9.1 lists the available render functions beyond `renderPlot()` for visualizations. Other possibilities for generating interactive output include interactive text, images, and tables.

Function	Output
renderPlot	Plot Output
renderText	Text Output
renderPrint	Printable Output
renderDataTable	Table output with DataTables JavaScript library
renderImage	Image file output
renderTable	Table Output
renderUI	UI Output
downloadHandler	File Downloads

Table 9.1: Shiny render functions from which to create a Shiny UI component for output.

library function, Section 1.1.2, p. 6

Listing 9.1 shows that the `server.R` file contains more than the `server()` function. As with any R code, invoke the `library()` function to access any functions that appear from contributed packages, here both the **shiny** package, and the **lessR** package to access the `Read()` and `Histogram()` functions. The server code is where the data gets processed, so the data must be available.

Read function, Section 1.2.4, p. 12

Shiny user-interface

The Shiny user interface generates a web page to receive the user-specified input, send to the server, receive the server-generated output, and then display the server output. Because the *ui* generates a web page it writes code in the standard web page languages of HTML, css, and JavaScript. Fortunately, Shiny takes care of all the technical web details, so the Shiny app developer only need know basic R with the Shiny enhancements, though certain customizations can be added with some basic knowledge of web page construction.

user interface, ui: The regions for the display of input and output and the R code that generates these regions.

The user only requires a web browser to interact with a Shiny app, free from knowledge of either R or web page construction. Shiny offers R analysis without the need to code in R. Employees of a company can examine various aspects of sales data, such as by region, without coding. Statistics students can focus on concepts instead of programming.

The Shiny *user interface* consists of regions of input and output on the generated web page, as well as other information such as a title. Shiny refers to the code that generates this interface as output from a Shiny function named, *ui*. The corresponding `ui.R` code that pairs with the `server.R` code from Listing 9.1 appears in Listing 9.2.

```
1   library(shiny)
2
3   ui <- fluidPage(   # Define UI for app that draws a histogram
4
5     titlePanel("Select Bin Width"),   # App title
6
7     sidebarLayout(   # Sidebar layout with input and output definitions
8
9       sidebarPanel(   # Sidebar panel for inputs
10
11        sliderInput(inputId = "myWidth",   # Input: Slider for the number of bins
12                    label = "Bin Width:",
13                    min = 1000, max = 20000, value = 8000)
14      ),
15
16      mainPanel(   # Main panel for displaying outputs
17        plotOutput(outputId = "myPlot")   # Output: Histogram
18      )
19    )
20  )
```

Listing 9.2: Example of `ui.R` code that defines an *ui* object to provide input for histogram bin width and then display the resulting histogram generated by the `server.R` code.

The core of the `ui.R` code is the `sliderInput()` function defined in Lines 11-13 of Listing 9.2. This function generates the web code that displays the slider input widget for entering bin width. The key parameter is *inputId*, listed first. It is here that the name of the input object is defined, specified as *myWidth* in this example, with the full name of *input$myWidth*. This value is then passed to the server.

The *label()* function provides an explanatory label over the widget to guide user input. Different widgets can have different parameters. For the slider, the specified parameters are *min* and *max*, for the minimum and maximum values displayed, and *value* for the default value of the initial bin width for the initially generated histogram before moving the slider to input custom values.

Table 9.2 lists the standard input widgets beyond the `sliderInput` function. All the Shiny input functions that define the widgets have the same general syntax. The arguments to each input function begins with *inputID* and *label*.

Function	Widget	inputId, label + Common Args
actionButton	Action Button	
checkboxInput	Single check box	value
checkboxGroupInput	Group of check boxes	choices, selected
dateInput	Calendar to aid date selection	value, min, max, format
dateRangeInput	date range from two calendars	start, end, min, max, format
fileInput	Upload file	multiple
numericInput	Field to enter numbers	value, min, max, step
radioButtons	Set of radio buttons	choices, selected
selectInput	Box with choices to select from	choices, selected, multiple
sliderInput	Slider bar	min, max, value, step
textInput	Field to enter text	value
submitButton	Submit button	text (no inputId or label

Table 9.2: Shiny widgets for which to input information.

panel: A specific region of the user interface.

Shiny refers to the physical regions of the web page user interface as *panels*. The typical Shiny visualization includes code for three panels: Title for the input panel, the input panel, and the output panel. The `sidebar()` function that begins on Line 9 specifies the layout of the input widget, here placed to the side of the web page, space permitting.

The `mainPanel()` function defines the panel that presents the output. Here only a single plot is presented, specified by the `plotOutput()` function. The key parameter is *outputId*, the name of which matches the name of the output from the corresponding render function defined in `server.R`, with full name *output$myPlot* in this example.

Define both the `sidebar()` and `mainPanel()` functions within the scope of the `sidebarLayout()` function that specifies the relative positioning of the input and output panels. The `titlePanel()` on Line 5 defines a title for the web page. Enclose all of these functions within the `fluidPage()` function that automatically adjusts the page display according to the available width. View the web page on any device that runs a web browser, from smartphone to large external computer monitor.

The output functions place different types of output into your Shiny app, such as plots, tables, and text. The first argument to an output function is *outputId*. Many render functions have corresponding output functions, such as `renderPlot()` and `plotOutput()`, as shown in Table 9.3.

Function	Output
dataTableOutput()	interactive table
htmlOutput	raw HTML
imageOutput	image
plotOutput	plot
tableOutput	table
textOutput	texzt
uiOutput	Shiny UI element
verbatimTextOutput	text

Table 9.3: Shiny output functions from which to output information.

9.2   Running a Shiny App

9.2.1   Shiny within RStudio

Each Shiny app, which consists of the `ui.R` and `shiny.R` files, or a single `app.R` file, is best contained within a single, unique RStudio project. To initiate an RStudio project, move the mouse to the top-right corner of the RStudio window and activate the New Project option in the drop-down menu as shown in Figure 9.2.

Figure 9.2: New project option in the drop-down menu at the top-right of the RStudio window.

Then choose to either create a new directory or use an existing directory. The name of the directory is the name of your created project, as in Figure 9.3.

Figure 9.3: Choose an existing directory or create a new directory with the name of your project.

The RStudio project directory is the place to store your `ui.R` and `server.R` files, as well as any needed data files. Here follow the recommended practice of storing data files in the *data* directory. View the directory contents from your file system, or, as shown in Figure 9.4, from the Files tab in RStudio, by default in the bottom right pane. The directory also includes the .Rproj file that defines your project. Double-clicking on this file opens the project in RStudio.

To run the interactive visualization, open the `ui.R` and/or `server.R` files by double-clicking on their name in the RStudio Files tab or your file system. Recognizing the respective file names, RStudio is aware of the Shiny app, and so displays a Run

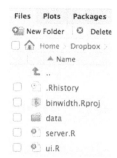

Figure 9.4: Contents of a RStudio project directory.

App button at the top of the Source window, by default the top-left window panel, shown in Figure 9.5.

Figure 9.5: Run an interactive visualization.

Click the Run App button to create the interactive visualization. The result appears in Figure 9.6.

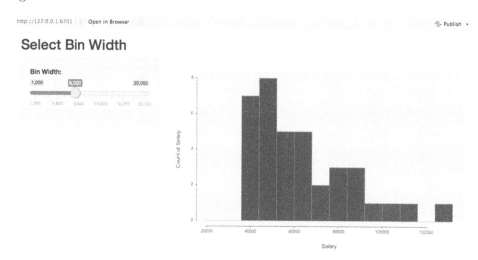

Figure 9.6: The web page for input/output.

9.2.2   Publish Shiny Apps on the Web

When the Shiny app works locally on your computer, publish it to the web for others to access. One possibility is to install the Shiny server software on a server computer connected to a local network and perhaps the Internet. Another possibility is to access a commercial server with Shiny server software installed, such as provided by RStudio, Inc.

1. Sign up for a shinyapps.io account that is free for limited use.

2. Follow the instructions when registering for an account for connecting your installation of RStudio to your shinyapps.io account. Instructions are also available here.

3. Run the Shiny app locally on your computer and press the Publish button at the top-right of the browser-like window that RSstudio opens. Provide a sensible title for your app and then click publish.

4. Check whether the deployed app runs the same on your local computer. If the app does not display correctly, examine the log files for the app.

5. To connect your locally developed Shiny app to the Shiny server at shinyapps.io, use the `deployApp()` function from the `rsconnect` package.

To connect Shiny applications on your computer with the Shiny app web hosting at `shinyapps.io` requires a hosting account. For your local computer to connect your app to the Shiny server requires the `rsconnect` package. Without a specified path name in the call to its `deployApp()` function, the reference is from the current working directory, such as from within an RStudio project.

R Input *Connect local Shiny apps to the Shiny server at shinyapps.io*

```
library(rsconnect)
deployApp()
```

Once you have your account and a shiny app uploaded to your account, ready to share with the world. Enter a URL of the following form to access:

`https://user_name.shinyapps.io/app_name/`

If other users access your app, you may run out of the allotted time for a free account. One way to conserve time is to decrease the idle time of an app from the default of 15 minutes to 5 minutes or so. To do so, from your shinyapps.io dashboard, choose Settings and then, under the heading of Instance, lower the setting for Instance Idle Time.

R Input *Reference*

functions `https://shiny.rstudio.com/reference/shiny/latest/`
summary `https://www.rstudio.com/wp-content/uploads/2015/02/shiny-cheatsheet.pdf`
video `https://vimeo.com/rstudioinc/review/131218530/212d8a5a7a/#t=37m46s`

Of course, there is much more to Shiny that presented in this introduction. This brief introduction to Shiny, however, provides the needed background to begin to develop and host Shiny apps. More sophisticated applications build on this foundation. Moreover, other R packages exist for interactive visualizations. One such package, `plotly` (Sievert, 2018), also has versions for the languages Python and MATLAB, and integrates well with `ggplot2` and Shiny apps. The concept of a widget is formalized in the `htmlwidgets` package (Vaidyanathan, Xie, Allaire, Cheng, & Russell, 2018), which allows custom extensions to Shiny apps as well as interactive enhancements to R Markdown documents. Interactive network visualizations are possible with the `visNetwork` package (Almende B.V., Thieurmel, & Robert, 2018), that also integrate with Shiny apps.

Chapter 10

Customize Visualizations

A data visualization portrays only one of many possibilities for each visual aesthetic. Customize virtually any aspect of a visualization such as the width of plotted lines, font sizes, and color. Organize these customizations according to both persistence and scope. Figure 10.1 illustrates the four possibilities.

Temporal

	Current Visualization	Persistant
General Theme	Change theme only for current visualization	Change theme indefinitely
Individual Characteristic	Change one or more individual aesthetics only for current visualization	Change one or more individual aesthetics indefinitely

Scope

Figure 10.1: Four types of customizations.

style modification, temporary: Applies to only a single visualization.

style modification, persistent: Applies to all subsequent visualizations unless explicitly modified.

A *temporary* modification applies only to a single visualization. A *persistent* modification applies to all subsequent visualizations unless subsequently modified. Regarding *scope*, the modification can apply to a single characteristic of the visualization or to an entire set of characteristics that together define a theme.

Customizing a visualization often includes the use of color, explored next.

10.1 Color References

10.1.1 Describe Colors

R provides multiple ways to specify a color, including RGB and HCL color spaces. First, however, consider color names.

showColors(), lessR: Show all named R colors and corresponding RGB components.

Color Names

R provides its full list 657 color names from its `colors()` function (that includes different spellings for the same color). Some of the more intriguing color names include `"chocolate"`, `thistle`, `"lavenderblush"`, and `"peachpuff"`.

To view the colors invoke the `lessR` function `showColors()`, which generates a `pdf` file that contains all the color names and also a sample of each color with its corresponding red, green, and blue composition. Figure 10.2 illustrates a small excerpt of the output of `showColors()` for some grayscale colors.

gray58	**gray59**	**gray60**	**gray61**
148, 148, 148	150, 150, 150	153, 153, 153	156, 156, 156

Figure 10.2: Some grayscale colors excerpted from the output of `showColors()`.

Grayscale varies from black, "gray00", to white, "gray100". These color names reference the underlying function gray() with input values from 0 for black to 1 for white, such as gray(0) and gray(1), with gray(.5) a gray of middle brightness.

Color names are simple to use, but ultimately more control over the specific colors chosen for a visualization is needed. That is the purpose of a color space.

RGB color space

The rgb() function provides a means to specify a color according to the red, green and blue components of light. An optional fourth component, alpha, can be added to achieve partial transparency. By default, the three or four values of the rgb() function are proportions, which vary from 0 to 1. For example, rgb(0,0,0) is equivalent to the named color "black", rgb(1,1,1) is equivalent to the named color "white", and rgb(1,0,0) is equivalent to the named color "red". The default alpha value of 1 results in an opaque color, that is, no transparency. Other applications for color manipulation, such as Adobe Illustrator for drawing and Adobe Photoshop for photographs, specify colors with values that range from 0 to 255. To apply this larger range of numbers, add the parameter and value maxColorValue=255 to the rgb() function call.

rgb() function, base R: Define a color in terms of red, green, and blue components.

R processes colors in standard hexadecimal notation, so R evaluates rgb() to hexadecimal. The hexadecimal digits vary from 0 through 9 then A through F to span 16 digits, equivalent to our usual base 10 numbers, but from 0 to 15. The first two digits of a color's hexadecimal representation indicate the amount of red, the second two digits indicate green, and the last two digits indicate blue. Both notational systems refer to the same color space, the underlying color expressed in terms of the amount of the red, green and blue components. An example of the evaluation of blue expressed in RGB notation follows in Listing 10.1.

```
> rgb(0,0,1)
[1] "#0000FF"
```

Listing 10.1: Hexadecimal output of the base R rgb() function.

In this example, the red and green components of the rgb() function call are both 0 with blue at the maximum value of 1. This translates into the hexadecimal equivalent of 00 for red, 00 for green, and the largest hexadecimal value expressed as FF for blue.

The rgb() specification follows from the physical characteristics of how a computer screen displays color. Unfortunately, engineering specifications of physical hardware do not translate into uniform human perception of colors. For example, there is no straightforward way to systematically vary colors and maintain the same levels of brightness with RGB colors. How we perceive color and how computer monitors work are two different processes, which suggests the need for another color space.

hcl() function, base
R: Define a color in
terms of hue, chroma,
and luminance.

HCL color space

The generally preferred specification of colors for data visualizations, the default for
both `lessR` and `ggplot2`, follows from the three HCL coordinates from the base R
`hcl()` function. The primary advantage of this color space is that different regions
of a visualization plotted in different hues can maintain the same level of intensity.
Otherwise, more intense, brighter regions are visually favored over darker regions.

The HCL color space defined by these coordinates expresses colors as we perceive
color, in terms of *hue* (the color name as positioned on a rainbow), *chroma* (intensity
of color from gray to colorful), and *luminance* (brightness). Express the hues of the
HCL color space in terms of degrees from 0 to 360 (and multiples of) with red set as
0 degrees. The other primary colors of light, green and blue, are set at 120-degree
increments, 120 and 240 degrees, respectively. With names selected by your author,
Figure 10.3 names the HCL hues around the color wheel in 30-degree increments
identified by the `lessR` color management system.

*visualize the HCL
color wheel with
colors*, Section 10.2.1,
p. 215

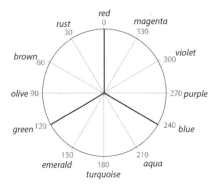

Figure 10.3: HCL named hues in 30-degree increments.

The parameter names for the three `hcl()` coordinates are `h`, `c` and `l`. The `hcl()`
function scales chroma and luminance dimensions from 0 to 100. Regardless of the
value of `h`, a value of 0 for `c` results in grayscale, from dark colors of gray for values
of `l` close to 0 and lighter colors of gray as the value of `l` approaches 100. Increasing
`c` closer to 100 yields highly saturated, more vivid colors. Values of `l` close to 100
result in pastels. Low values of `l` result in deeper, more vibrant colors.

As with the output of `rgb()`, the R visualization functions allow customization
with direct entry of the hexadecimal output of `hcl()`, an example of which appears
in Listing 10.2. Not all of its colors of the HCL color space of human perception,
unfortunately, translate into the RGB color space of computer monitors. To display
all colors from `hcl()` with valid input values requires approximation of some of the
colors translated to the RGB color space (and hence the equivalent hexadecimal).
The `hcl()` function parameter `fixup` defaults to `TRUE`, which permits approximation
of colors as needed. Set the parameter to `FALSE` to display only true HCL colors. The
problem is sufficiently pervasive that most combinations of chroma and luminance
do not yield the full spectrum of HCL colors. Only moderate values, such as `c` set
to 50 and `l` set to 65, render all hues as proper HCL.

```
> hcl(240,70,50)
[1] "#007FBB"
```

Listing 10.2: Output of the `hcl()` function for one color, a shade of blue.

10.1.2 Parameters `fill` and `color`

Given the ability to specify colors in R, how to customize colors of specified characteristics of a visualization? Two useful parameters for this task are `fill` and `color`. The parameter `fill` refers to the *interior* of a polygon, such as a point, bar, or area of a pie slice. The `fill` parameter describes how the polygon appears from the inside. The parameter `color` refers to the perception of color viewed from the *exterior* of a polygon or line, its appearance from the outside. For a polygon, `color` refers to the color of its borders. For line segments, `color` refers to the colors of the lines. Figure 10.4 illustrates the use of these two parameters by redrawing the default bar charts from Figure 2.1 with a lighter gray fill and black border for each bar.

`fill` parameter: Interior color of a polygon.

`color` parameter: Color of a line or border of a polygon.

default bar charts, Figure 2.1, p. 29

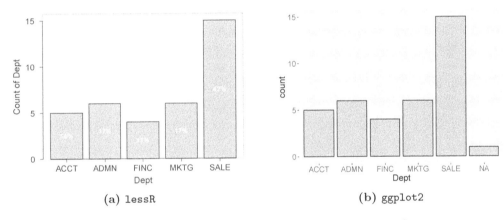

(a) `lessR`   (b) `ggplot2`

Figure 10.4: Bar chart with assigned fill and color values.

The code that customizes the `fill` and `color` parameters follows.

R Input *Basic bar chart with assigned fill and color values*
data: d <- Read("http://lessRstats.com/data/employee.csv")

lessR: BarChart(Dept, fill="gray20", color="black")
ggplot2: ggplot(d) + geom_bar(aes(Dept), fill="gray20", color="black")

In place of the color names `"gray20"` and `"white"`, any other valid R reference to colors could be substituted. These other possibilities include the use of other color names as well as via the functions `rgb()` or `hcl()`.

10.2 Palettes

color palette: A set
of related colors.

Visualizations apply combinations of related colors, including shades of gray for
grayscale. Examples include variations of a single hue, or multiple hues at the
same levels of grayscale saturation (chroma) and brightness (luminance). A set of
related colors forms a *color palette*, expressed as a vector of color names. Include a
corresponding palette to help communicate the associations and patterns revealed
by visualizations, either from a referenced global theme, perhaps the default theme,
or customize with an individually selected palette.

**perceptual
uniformity**: Plotted
colors change
appropriate to the
amount of change in
the corresponding
data values.

An optimal color palette satisfies several criteria. For a continuous variable, central
to the mapping of data values to perceived colors is perceptual uniformity. The
colors in a *perceptually uniform* palette appropriately change corresponding to the
changes in the magnitude of the data values. For example, the palette is *not* uniform
if a small change in data values leads to a large change in the displayed colors. A
perceptual non-uniform mapping of data to colors yields a distorted perception of
the corresponding changes in the underlying data values.

For categorical variables, with visualizations such as bars of a bar chart or slices
of a pie chart, then all the colors generally should be presented with the same
level of chroma and luminance so that some areas of color are not brighter than
others. Otherwise, for example, a bright red bar may draw the viewer's attention
to that area more so than an adjacent dark blue bar. The different colors should
distinguish different areas that correspond to different categories but not bias the
visual attention of some areas relative to others.

All visualization systems feature built-in color palettes. Users are also free to define
any palette of their choosing, either generated by specialized functions or manually
entered. Three basic types of color palettes appear in visualizations: qualitative,
sequential, and divergent.

10.2.1 Qualitative Palettes

qualitative scale:
Color palette of
mixed hues applied
to nominal data.

nominal data,
Section 1.2.3, p. 11

An example of a categorical variable with unordered levels is State of Residence
in the USA of the respondents to a survey. One state is not more or less than
another state, so State of Residence yields nominal data. One option displays all
levels – such as bars on a bar chart, slices of a pie chart or different colored points
depending on the level – with the same color. To apply a color palette, however,
the appropriate scale for the display of the levels of nominal data for a categorical
variable is a *qualitative scale*, sometimes referred to as a discrete scale. Each area
that corresponds to a specific level of the variable portrays its own hue, resulting in
a rainbow of colors. In general, specify the colors in the HCL color space derived
from the HCL color wheel shown in Figure 10.3, all with different values of hue but
the same chroma and luminance.

HCL equal chroma and luminance palettes

The palettes. The `lessR` function `getColors()` generates the colors for a wide variety of color palettes and/or displays the resulting colors, including the HCL color wheel from Figure 10.3. To do so with the `ggplot2` ecosystem requires two functions from the `scales` package. This package is not part of the tidyverse, so download separately and invoke the `library()` function to access its constituent functions.

getColors() function, `lessR`: Generate and display color names in a palette.

> **R Input** *Generate an HCL color qualitative color palette for 12 hues*

lessR: getColors(shape="wheel")
scales: show_col(hue_pal()(12))

library() function, Section 1.1.2, Page 6

The parameter list for `getColors()` begins with `pal`, for palette, with a default value of `"hues"`, which generates the specified number of hues around the HCL color wheel. The number of colors for the generated scale is `n` with a default value of 12. The function `getColors()` generates by default a rectangle divided into intervals of hues in the order they are invoked in subsequent visualizations. Adjacent values were chosen to maximize hue separation. The value of the hue is `h`, presented in the plot as well as the console with the hex and RGB information. To generate an actual color wheel, which corresponds to Figure 10.3, add the parameter `shape` set to the value of `"wheel"`.

Generate the vector of color hexadecimal codes with the `hue_pal()` function and display them in a rectangular grid with the `show_col()` function. Specify the number of colors as part of the function call, as illustrated below, here for 12 colors. The HCL hue values are not presented. Both `getColors()` and `hue_pal()` generate a vector of color names in hexadecimal notation, made explicit by saving the output of each function into a vector, here *clr*, shown in Listing 10.3. Because the output of the functions are vectors of color hexadecimal codes, `getColors()` can display colors from `hue_pal()` and `show_col()` can display colors from `getColors()`.

hue_pal() function, `scales`: Generate vectors of color names of varying hues from the HCL color space.

show_col() function, `scales`: Display colors from names as a rectangular grid.

```
> clr <- getColors()
> clr
 [1] "#2D8BC3" "#A57E08" "#51932E" "#C7657B" "#8E76C9" "#009B8B"
 [7] "#BB714D" "#838A00" "#0097AD" "#C561A2" "#009962" "#B068BE"

> clr <- hue_pal()(12)
> clr
 [1] "#F8766D" "#DE8C00" "#B79F00" "#7CAE00" "#00BA38" "#00C08B"
 [7] "#00BFC4" "#00B4F0" "#619CFF" "#C77CFF" "#F564E3" "#FF64B0"
```

Listing 10.3: Color name vectors.

The 12 colors differ from the two functions because `getColors()` applies default chroma and luminance at 65 and 55, respectively, whereas `hue_pal()` defaults these values to 100 and 65. The processing of these color specification vectors illustrates a fundamental design distinction between `lessR` and `ggplot2`. The former minimizes the number of functions, whereas the latter applies a larger number of functions, here one to generate the palette and another function for its display. Moreover, `lessR` uses `getColors()` to generate a variety of palettes, whereas the `ggplot2` ecosystem via the `scales` package provides a different function for each palette style.

To explore the visual properties of the HCL color space, consider the HCL color wheel for different values of chroma and luminance across the varying hues. High values of chroma and luminance generate pastel colors. Use c and l to specify these values, which are then passed on to the `hcl()` base R function to display the colors.

R Input *Generate HCL pastel colors*

```
lessR: getColors(n=8, c=100, l=90)
scales: show_col(hue_pal(c=100, l=90)(8))
```

Obtain vibrant, rich hues from high values of chroma but low values of luminance.

R Input *Generate HCL rich, vibrant colors*

```
lessR: getColors(n=8, c=90, l=40)
scales: show_col(hue_pal(c=90, l=40)(8))
```

With the HCL color space, holding *chroma* and *luminance* constant, only varying *hue*, the grayscale conversion of the resulting rainbow from any of these generated HCL color palettes provides a constant level of gray across the entire spectrum. The result is that some colors are not brighter or flashier than less saturated colors. The `hcl()` color space provides a more useful comparison of different colored regions in visualizations such as bar charts, maps, and continuous color graduations, without bias towards brighter colors.

Application of the palettes.   Both `lessR` and `ggplot2` present default qualitative scales. The distinction is that by default (that is, without first specifying `style("gray")`), `lessR` displays the default qualitative scale `"colors"` for bar charts and pie charts. The ordering of the hues was selected to maximize differences between adjacent hues, beginning with blue (240), brown (60), green (120), red (0), and purple (275) for the first five hues. Many palettes with `lessR` are pre-defined for easier access than with the full specification with `getColors()`, such as the default qualitative color palette.

map data values into a visual aesthetic: Color, size, or shape of a plotted object depends on another variable.

To display the default rainbow of colors for a bar chart with `ggplot2`, *map values* of the variable plotted to the color aesthetic with the `fill` parameter. The mapping assignment of color differs from setting `fill` to a color or a vector of colors, as in Figure 10.4. In this example, *map* the colors from data values as opposed to **set** their values, that is, assign the color of each bar according to the corresponding value of the `fill` variable, here *Dept*. Because *Dept* is categorical, the default qualitative scale is assigned to map the colors. The mapping of a variable to a visual aesthetic yields a legend, which scales the values of the mapped variable according to that aesthetic, such as color.

map vs. set visual aesthetics Displayed aesthetics depend on the value of another variable vs. independent of other variables.

For `ggplot2`, assign the `fill` parameter for mapping to a variable as part of `aes()`, which specifies the mapping of the data values to the specified visual aesthetic. In contrast, in Figure 10.4 the `fill` was set to a constant color *outside* of `aes()`.

set colors,
Figure 10.4, p. 213

> **R Input** *Bar chart with qualitative palette of hues*
>
> *data*: `d <- Read("http://lessRstats.com/data/employee.csv")`
>
> ---
>
> *lessR*: `BarChart(Dept)`
> *ggplot2*: `ggplot(d) + geom_bar(aes(Dept, fill=Dept))`

To change the default behavior of `lessR` to display all the bars in the same color, as does `ggplot2` by default, either set the color with `fill` or separately invoke the `style()` function to choose another theme, such as `"lightbronze"`. For the default `"colors"` style, `lessR` implicitly maps the qualitative scale of different hues to the bars, but for other styles sets the color of all the bars the same according to the color of the current style.

The colors in the qualitative palette can be customized. Here select richer, more vibrant colors with increased saturation to c=90 and decreased luminance to l=40. To customize with `lessR`, call `getColors()` with the custom values of c and l. With `ggplot2` specify the custom values with the `scale_fill_hue()` function, which calls the `hue_pal()` function that generates the colors.

scale_fill_hue()
function, `ggplot2`:
Generates the default
color palettes for
categorical variables.

> **R Input** *Bar chart of counts with custom colors*
>
> *data*: `d <- Read("http://lessRstats.com/data/employee.csv")`
>
> ---
>
> *lessR*: `BarChart(Dept, fill=getColors(c=90, l=40))`
> *ggplot2*: `ggplot(d) + geom_bar(aes(Dept, fill=Dept)) +`
> `        scale_fill_hue(c=90, l=40, na.value="black")`

In general, follow the principle that the colors of a qualitative palette should share the same levels of chroma and brightness. Other possibilities, however, can also be considered to achieve specific effects.

Wes Anderson palettes

Another criterion for generating a palette is aesthetic design. The movie director Wes Anderson creates movies with engaging color themes, favoring pastels intermixed with bold primary colors. Karthik Ram and Hadley Wickham (2018) developed a package of palettes, `wesanderson`, created from the color theme of many of his movies. The available themes: `"BottleRocket1"`, `"BottleRocket2"`, `"Rushmore1"`, `"Royal1"`, `"Royal2"`, `"Zissou1"`, `"Darjeeling1"`, `"Darjeeling2"`, `"Chevalier1"`, `"FantasticFox1"`,`"Moonrise1"`, `"Moonrise2"`, `"Moonrise3"`, `"Cavalcanti1"`, `"GrandBudapest1"`, `"GrandBudapest2"`, `"IsleofDogs1"`, and `"IsleofDogs2"`.

`lessR` incorporates these themes, directly accessed from the `fill` and `color` parameters. To plot with `ggplot2`, first access the `wesanderson` package, then generate the vector of color names with the `wes_palette` function. Specify the `ggplot2` function

scale_fill_manual():
Provide an arbitrary
vector of color names
to provide a palette
for visualization.

`scale_fill_manual()`, the function that sets the visual aesthetics according to the specified vector of color names.

The following examples color the bars of a bar chart according to the palette "Royal1" based on Wes Anderson's movie The Royal Tenenbaums.

> **R Input** *Filled bars in a bar chart from a wesanderson palette*
>
> *data*: d <- Read("http://lessRstats.com/data/employee.csv")
>
> ---
>
> *lessR*: BarChart(Dept, fill="Royal1")
> *wesanderson*: my.palette <- wes_palette("Royal1", n=5, type="continuous")
> *ggplot2*: ggplot(d) + geom_bar(aes(Dept, fill=Dept)) +
>   scale_fill_manual(values=my.palette)

All of the themes from the **wesanderson** package such as "Royal1" only define four or five distinct colors. The `wes_palette()` parameter "type" has a "continuous" option, which interpolates intermediate colors to provide for more possibilities, such as the five colors needed for the five levels of the *Dept* variable. **lessR** automatically sets `type` to continuous and provides the corresponding number of intervals for which to provide colors.

10.2.2   Sequential Palettes

sequential palette:
Ordered progression
of colors for the same
hue.

A *sequential palette* applies to a variable with ordered values. The qualitative scales vary hue. One strategy for a continuous variable retains the same hue but varies luminance, or possibly chroma, or both. Or, specify a range of hues in a progression. Both **lessR** and **ggplot2** also recognize an ordered representation of the levels of a factor, which demonstrates underlying continuity with sequential scales.

Constant hue palettes

The palettes. To generate a sequential scale of HCL colors set the hue, `h`, set the value of chroma, `c`, and then specify a two-valued vector for luminance, `l`, which indicates the start and end values of luminance across the range of `n` intervals. To obtain a grayscale, set chroma to 0 and vary luminance, illustrated for eight intervals in Figure 10.5.

lessR relies upon `getColors()` to create a sequential palette, including grayscale. Set `c` to 0 and provide a lower and upper value for luminance. The **scales** package, which provides the scales referenced by **ggplot2**, provides a specialized function, `show_col(gray_pal)`, to generate a grayscale. Specify the grayscale range here by the gray value on the 0 1o 100 scale.

> **R Input** *Generate, display grayscale*
>
> ---
>
> *lessR*: getColors(n=8, c=0, l=c(15,94))
> *scales*: show_col(gray_pal(start=0.2, end=0.8)(8))

"blues", **"reds"**,
etc: 12 pre-defined
lessR sequential
scales according to
the names of 12
corresponding HCL
colors.

To further simplify, `getColors()` provides pre-defined sequential ranges based on the corresponding hue. The names of the pre-defined ranges are the hue names from

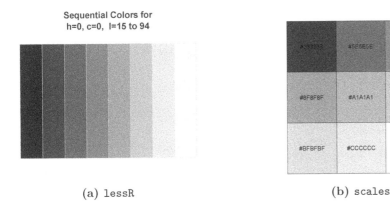

Sequential Colors for
h=0, c=0, l=15 to 94

(a) lessR   (b) scales

Figure 10.5: Grayscale palettes.

the color wheel in Figure 10.2.1 with an added "s". For example, the hue at 240 degrees, `"blue"`, defines a corresponding sequential range named `"blues"`. Ranges are defined for each of the 12 hue names plus `"colors"`, the default, for qualitative scales, and `"grays"` for grayscale. The function generates scales for the specified value of n and adjusts the extent of the range of luminance according to the number of intervals generated, so the fewer intervals the less the range.

The following simpler function call generates the same `lessR` grayscale in Figure 10.2. Custom values of the start and end points of luminance can also be specified with the l parameter.

color wheel names,
Section 10.2.1, p. 216

```
lessR: getColors("grays", n=8)
```

Substitute `"blues"` or `"emeralds"` or `"purples"` or any other plural version of the HCL color wheel names from Figure 10.2.1 to access a corresponding pre-defined sequential color scale, illustrated next. For the `scales` package the function to generate a color sequential range is `seq_gradient_pal()`. Unlike the corresponding hue and grayscale functions that generated a discrete number of intervals, this function operates over a continuous range of values from 0 to 1.

R Input *Generate, display a sequential scale of blues, varying chroma and luminance*

```
lessR: getColors("blues", n=12)
scales: x <- seq(0, 1, length.out=12)
        getColors(seq_gradient_pal(low=hcl(240,35,94),
                high=hcl(240,75,15))(x))
```

The pre-defined palettes such as `"blues"` pre-select the hue as well as the chroma and luminance. Include custom values for `c` and `l` in the call to `getColors()` to override the provided default values. The default level of chroma for each pre-defined palette varies from 35 to 75. The default range of luminance depends on the number of intervals generated.

`lessR` also provides a function, `showPalettes()`, that by default presents the 12 pre-defined sequential palettes, the qualitative HCL palette, and the grayscale palette.

The default number of intervals is 12, but can be adjusted as desired. The default value for the `pal` parameter to generate these palettes is `"hues"`. Other possible values are `"wesanderson"`, and `"viridis"`, discussed next.

The viridis palettes

The `viridis` palette (Rudis, Ross, & Garnier, 2018) was developed according to the following principles.[1]

 ▷ Span as wide a color palette as possible to accentuate differences in adjacent colors, with outliers readily detected

 ▷ Perceptual uniformity, so that data values similar in value have similar-appearing colors, and data with discrepant values have more different-appearing colors, consistently across the range of values

 ▷ Robust to colorblindness, so that the above properties hold true for people with common forms of colorblindness, as well as in grayscale printing

Both `lessR` and `ggplot2` incorporate direct access to the family of `viridis` palettes, which includes `viridis` plus the `magma`, `plasma`, `inferno` and `cividis` palettes. To view the family of five `viridis` palettes in the `viridis` and `viridisLite` packages, call the `lessR` function `showPalettes()`. The default value for the number of intervals is 12, but specify in general with `n`. A large value such as `n=100` approximates continuity.

> *lessR*: `showPalettes("viridis", n=100)`

library(),
Section 1.1.2, p. 6

To view only the `viridis` palette use, again, `getColors()` with `lessR`. For `ggplot2`, use `viridis_pal()` from the related package `scales`, which is not loaded with the tidyverse packages, so first enter a separate `library()` call.

> *lessR*: `getColors("viridis")`
> *scales*: `show_col(viridis_pal()(12))`

The `viridis` palette spans a wide color gamut, from dark blue to bright yellow. One implication is that extreme data values, large or small, shown as either dark blue or bright yellow, are differentiated from the other data values. In terms of perceptual uniformity, no large color steps intrude across adjacent color values. Instead, dark blue smoothly transitions from to blue to blue-green to green to yellow. The most common form of colorblindness does not distinguish red from green, a problem avoided by the `viridis` palette, which is free of red hues.

To view other palettes in the `viridis` family, for `lessR` substitute the name of the other palette, such as `"magma"` for `"viridis"` in the call to `getColors()`. For `ggplot2`, specify the same name in quotes, but as the value for the `option` parameter in the call to `show_col(viridis_pal()`. The `viridis_pal()` function also offers

---

[1]Stefan van der Walt and Nathaniel Smith (2015) developed the `viridis` and related continuous color palettes for the Python language. Simon Garnier and others (2018) ported `viridis` to R into two packages, `viridis` and `viridisLite`.

parameters `start` and `end`, scaled on the 0 to 1 interval for viewing a subset of the full palette.

As a sequential palette, any palette of the `viridis` family is usually best applied to a variable with ordered levels of magnitude, ordinal variables and numerical variables from a continuum. As discussed, R represents ordinal data with the data type of an ordered factor, an ordered progression of categorical data values such as the values of the variable *JobSat* of low, medium, and high. The ordered factor can be visualized as the primary aspect of the visualization, as applied to the bar chart, or it can serve as a grouping variable, such as plotting points in a scatterplot from different groups in different colors.

ordinal variable,
Section 1.2.3, p. 10

ordered factor,
Section 1.2.6, p. 21

Instead of `"viridis"` as the `ggplot2` default display for variables with ordered data values, for an ordered factor `lessR` defaults to a sequential palette with constant hue. The yellow color at one end of `viridis` scale is bright, but does not prominently show on a standard color monitor. The result, in the corresponding bar chart, for example, is that the yellow bar, in your author's opinion, appears too distinct from the remaining bars, and so is not subjectively perceived as part of the same continuum as the remaining bars. Or, with a scatterplot with points from different groups of an ordered variable displayed in different viridis colors, the points displayed in yellow are almost not recognizable, too faded, particularly against the default gray background of `ggplot2`.

*bar chart of an
ordinal variable,*
Figure 3.17, p. 65

scatterplot by groups,
Figure 5.9, p. 113

Easily access the `"viridis"` palette in `lessR` via, for example, the `fill` parameter. With `ggplot2`, first convert the variable to an ordered factor.

> **R Input** *viridis bar chart palettes scaled on x*
>
> *data*: d <- Read("http://lessRstats.com/data/Mach4.csv")
> ---
> *lessR*: BarChart(m06, fill="viridis")
> *base R*: d$m06 <- factor(d$m06, ordered=TRUE)
> *ggplot2*: ggplot(d, aes(m06, fill=m06)) + geom_bar()

Another example of obtaining a viridis scaling is based not on the value of the x or categorical variable, but on the value of y or numerical variable.

*map y to fill
aesthetic,* Figure 3.18,
p. 66

10.2.3 Divergent Palettes

Two colors anchor a divergent scale, one on each side, so that each fade to a neutral color toward the middle. The first example of divergent color scales is found in Figure 2.3, which consists of 20 one-column stacked bar charts. Each bar chart displays the responses to an item on a 6-pt Likert scale, scored with the integers 0 through 5 to represent the continuum from Strongly Disagree to Strongly Agree. The presentation in Figure 2.3 is limited to grayscale, but in practice, each side of the scale displays different, usually complementary, colors.

divergent palette:
the middle lighter
colors diverge on
each side to darker
colors.

*divergent color scale
for bar charts,*
Figure 2.3, p. 33

Figure 10.6 shows the divergent palette applied to a single bar chart in the traditional format. Each bar corresponds to the frequency of data values for a corresponding level of a categorical variable. Again, the bar chart anchors both sides with gray instead of two opposing colors.

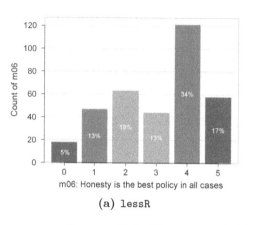

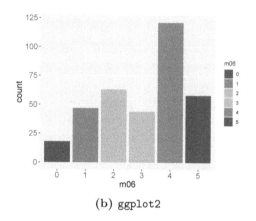

(a) `lessR`   (b) `ggplot2`

Figure 10.6: Divergent scaled bar charts of 6-pt Likert responses.

HCL color wheel,
Figure 10.3, p. 212

vectors, Section 1.2,
p. 6

For `lessR`, the most general method to define a divergent scale follows from the 12 pre-defined sequential palettes such as `"blues"` and `"greens"` based on the 12 HCL color names from Figure 10.3. The value of the `fill` parameter to define a divergent scale is a vector with two values. The first argument names the hue of the left side of the scale. The second argument names the hue of the right side.

example of `scale_`
`fill_gradient2()`,
ggplot2, Figure 10.6,
p. 222

The divergent `ggplot2` function `scale_fill_gradient2()` does not work in this situation because the scaling function applies only to continuous variables, yet `ggplot2` interprets the six integer responses as discrete. (See an example in the next chapter that illustrates the direct use of `scale_fill_gradient2()`.) However, `ggplot2` provides access to the core scaling function upon which `scale_fill_gradient2()` relies, `div_gradient_pal()` from the `scales` package. From the base R sequence function, `seq()`, generate six evenly spread numerical values from 0 to 1. From these values generate six discrete but diverging color names to enter into `scale_fill_manual()` to pass those six values to the plotting function.

R Input *Diverging scale for bar chart of Counts of Likert scale responses*

data: `d <- Read("http://lessRstats.com/data/Mach4.csv")`

     `d <- factors(m01:m20, levels=0:5, ordered=TRUE)`

lessR: `BarChart(m06, fill=c("grays", "grays"))`
scales: `clr <- div_gradient_pal(low="gray20", high="gray20")`
     `(seq(0,1,length.out=6))`
ggplot2: `ggplot(d, aes(m06, fill=m06)) + geom_bar() +`
       `scale_fill_manual(values=clr)`

For purposes of grayscale printing, define the divergent scale with the `"grays"` displayed on both sides of the palette. More generally, both colors would differ, usually to maximize hue differentiation about 180 degrees apart on the HCL color wheel from Figure 10.3. An example would be from `hcl(0,90,40)` to `hcl(240,90,40)`. For `lessR`, to get a neutral point on a true divergent scale, use of one the pre-defined scales such as `"reds"` to `"blues"`.

10.3   Themes

The values for each visual aesthetic, color or style option, can be pre-set as a group to define a *theme*, a presumably harmonious blend of visual aesthetics. Both `lessR` and `ggplot2` offer a variety of pre-defined themes, including their default themes. Both systems also allow the user to construct customized themes and then save for later use.

theme: Pre-defined values for all visual aesthetics that apply to all visualizations when activated.

The default `lessR` theme, `"colors"`, displays a relatively colorful palette. The `lessR` themes present different color combinations. Change or tweak a `lessR` theme with the `style()` function. The `theme` argument is the first argument in the parameter list for the function, so the first unnamed argument to `style()` specifies the theme. The `lessR` theme names correspond to the predominant theme color. Beyond the default of `"colors"`, lessR themes include `"lightbronze"`, `"dodgerblue"`, `"darkred"`, `"gold"`, `"darkgreen"`, `"blue"`, `"red"`, `"rose"`, `"green"`, `"purple"`, `"sienna"`, `"brown"`, `"orange"`, `"white"`, and, illustrated throughout this book, `"gray"` for grayscale, set with `style("gray")`.

style() function, **lessR**: Specify the general theme and/or modify individual characteristics for subsequent visualizations.

For additional flexibility, `lessR` also provides the `sub_theme` parameter. For example, a form of a grayscale visualization has white lettering on a black background. The sub-theme `"black"` sets a black background for any of the primary themes. To obtain grayscale with a black background, specify the `"black"` sub-theme with the `"gray"` theme. Other sub-themes change characteristics other than color as shown in Figure 10.7.

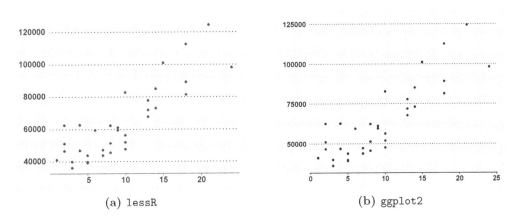

(a) `lessR`   (b) `ggplot2`

Figure 10.7: Implementation of the `wsj` theme from package `ggthemes` and the similar sub-theme from `lessR`.

The default `ggplot2` theme, `theme_gray()`, presents a gray background with white grid lines, as illustrated in the previous `ggplot2` visualizations. In addition to the default, themes include `theme_bw()`, `theme_linedraw()`, `theme_light()`, `theme_dark()`, `theme_minimal()`, `theme_classic()`, and `theme_void()`. Enter `?theme_gray` to see a brief description of all the themes.

The `ggthemes` package (Arnold, 2018) provides additional `ggplot2` themes, as well as additional `geoms` and `scales`. For example, there is `theme_wsj()` to produce plots

ggthemes package: Additional themes, scales, and geoms for `ggplot2`.

similar to the style used by the *The Wall Street Journal*, illustrated in Figure 10.7b. Another two themes, `theme_solarized()` and `theme_solarized_2()` produce plots based on solarized colors, a pleasant, precisely defined color palate for computer work (Schoonover, 2017). To access `ggthemes` themes, as with any R contributed package, during a working R session first install the package on your computer and then load the package with the `library()` function.

Figure 10.7 shows the `theme_wsj()` applied to the `ggplot2` scatterplot, and the corresponding `lessR` sub-theme, inspired by the `ggthemes` version. Unless creating the visualization in grayscale, the default background is a light bronze color. The modification in `lessR` is a sub-theme because it modifies the basic theme, applying the light bronze background and changing the format of the axes and grid lines, but retains the color of the plotted points according to the current theme.

install a package, Section 1.1.2, p. 5

library() function, Section 1.1.2, p. 6

10.3.1　Persistent Theme

theme_set() function, **ggplot2**: Specify the general theme for subsequent visualizations.

Individual customizations, Section 10.4, p. 225

Switch themes with a single function call, which persist across subsequent visualizations unless otherwise modified. With `lessR` specify the main color theme as `"gray"` to obtain grayscale, then, as illustrated below, further modify with the `wsj` sub-theme to conform to the `wsj` styling, such as the removal of the *y*-axis. For `ggplot2`, change the theme with the `theme_set()` function. Convert to grayscale by modifying the window background color to gray, as shown in a later section.

> **R Input** *Change themes*
>
> *lessR*: `style(sub_theme="wsj")`
> *ggplot2*: `theme_set(theme_wsj())`

10.3.2　Theme Applied to Current Visualization

To change the theme for a single visualization, modify the instructions to generate the visualization. In this `lessR` example, modify the appearance of the generated bar chart according to the `"darkred"` theme. Modify the resulting `ggplot2` bar chart according to the `classic` theme which presents a white background without grid lines.

> **R Input** *Bar chart of Counts with customized theme*
> *data*: `d <- Read("http://lessRstats.com/data/employee.csv")`
>
> *lessR*: `BarChart(Dept, theme="darkred")`
> *ggplot2*: `ggplot(d, aes(Dept)) + geom_bar() + theme_classic()`

For `lessR`, invoke the `theme` parameter to specify a theme different from the current theme, either the default `"colors"`, or set by the `style()` function. For `ggplot2`, modify a single visualization by adding the function for the theme to the functions that define the visualization.

10.4 Customize Individual Characteristics

Several examples throughout this chapter illustrate modifying the interior of bars and related polygons with the `fill` parameter and the edges of the polygons and lines with the `color` parameter. As seen, modify these colors either by mapping data values or setting colors. Apply these parameters directly to corresponding `lessR` and `ggplot2` visualization functions.

Many aspects of a visualization do not directly depend on data values, such as axis labels. By default `lessR` and `ggplot2` display the variable name as the axis label. If variable labels are present, however, `lessR` displays the variable name followed by the variable label.

variable labels,
Section 1.2.5, p. 18

To change the axis labels, `lessR` uses the standard base R parameters `xlab` and `ylab`, `sub` for a sub-title under the *x*-axis label, and `main` for the axis title. One `ggplot2` function to specify axis labels, `labs()`, includes parameters *x*, *y*, `title`, and `subtitle`. `ggplot2` also includes a parallel set of functions, `xlab()`, `ylab()`, and `ggtitle()` to accomplish the same purpose.

`labs` function,
ggplot2: Specify the content of axis labels.

> **R Input** *Bar chart of Counts with specified x-axis label*
>
> *data*: d <- Read("http://lessRstats.com/data/employee.csv")
> ---
> *lessR*: BarChart(Dept, xlab="Annual Salary (USD)")
> *ggplot2*: ggplot(d, aes(Dept)) + geom_bar() +
>   labs(x="Annual Salary (USD)")

`lessR` also has the function `label()`, which displays an existing variable label for a variable regardless of the visualization system. If variable labels are present in the `l` data frame, in the format of a variable name in the first column and variable label in the second column for all or some of the variables, then the following yields the same *x*-axis labels as the previous example.

> *ggplot2*: ggplot(d, aes(Dept)) + geom_bar() + labs(x=label(Salary))

Potential modifications, however, extend much further.

10.4.1 List of Individual Characteristics

Want green axis labels, extra large? A purple background? Want to customize almost any aspect of your visualization? View the available color and style options for either `lessR` or `ggplot2` with simple function calls. The output is a long list of customizable characteristics. For either system, this list is the key to changing the corresponding characteristics of the resulting visualization. Each list identifies the specific characteristics by their names, and then the properties of that characteristic amenable to modification.

> **R Input** *View theme parameters and current settings*
>
> *lessR*: `style(show=TRUE)`
>
> *ggplot2*: `theme_get()`

To illustrate, specify a new background color for the plot area of a panel, the rectangle defined by the x and y coordinate axes. Here display only the information for `lessR` parameters involved with the panel background color. Following each color specification are four integers, the red, green, blue and transparency components of the specified color, respectively, each in the range of 0 to 255. For both `lessR` and `ggplot2`, `fill` refers to the interior color of an object, such as a rectangle. The parameter `color` refers to the color of line segments, either by themselves or the segments that comprise the border of a polygon, such as a rectangle.

```
                          ─── some lessR parameters ───
BACKGROUND
window.fill .. Window fill color ......... 247 242 230 255
panel_fill ... Panel fill color .......... 255 255 255 0
panel.color .. Panel border color ....... 222 217 205 255
panel.lwd .... Panel border line width .. 0.5
panel.lty .... Panel border line type ... solid
```

R line types: "solid", "dashed", "dotted", "dotdash", "longdash", or "twodash", plus "blank", also referred to as the integers 1-6 and 0 for "blank".

Both `lessR` and `ggplot2` provide the standard six base R line types, plus `"blank"` for no line. In the `lessR` list, `panel.lty` refers to the assigned line type for each panel in a plot, here `"solid"`.

Following is the similar excerpt from the `ggplot2` list of parameters.

```
                          ─── some ggplot2 parameters ───
$ panel.background      :List of 5
 ..$ fill          : chr "gray92"
 ..$ colour        : logi NA
 ..$ size          : NULL
 ..$ linetype      : NULL
 ..$ inherit.blank: logi TRUE
```

What does the value of `NULL` mean for the `size` (width) and `linetype` parameters? Both `lessR` and `ggplot2` define a hierarchy of properties. At the top of the hierarchy of `ggplot2` characteristics are the most general: `line`, `rect` and `text`, from which properties lower in the hierarchy inherit their values unless otherwise modified. These values of `NULL` imply that these values are inherited from parameters higher in the hierarchy. For `line` there is a `linetype` property of type `num` for numeric with a value of `1`, which refers to a solid line segment. For `size`, also of type `num`, the value is `0.5`, the default line width.

From the list of available properties that can be modified, apply modifications that persist until another update to modify a general theme, or only modify a single visualization.

10.4.2 Customize a Single Analysis

This example modifies the scatterplot from Figure 2.7 to display the panel background as `"slategray1"`, and the axis text with `"darkred"`.

R Input *Temporary change of style element (for a scatterplot)*

```
lessR: style(panel_fill="slategray1", axis_text_color="darkred")
       Plot(Years, Salary)
       style()
ggplot2: ggplot(d, aes(Years, Salary)) + geom_point() +
           theme(panel.background = element_rect(fill="slategray1")) +
           theme(axis.text = element_text(color="darkred"))
```

For `lessR`, the relevant parameters are `panel_fill` and `axis_text_color`. For any parameter not in the parameter list of the visualization function, such as these two parameters, modify with the `style()` function. Changes with `style()` persist, applicable until explicitly modified with another call to `style()`. Invoke a new theme to simultaneously reset all style parameters consistent with that theme. The simplest function call to re-initialize is `style()`, which resets all parameters to the default theme, the equivalent of `style("colors")`.

The `ggplot2` function `theme()` modifies individual non-data components of a theme as they apply to a single visualization. Examples of non-data components include grid lines and axes. From the default theme, the `fill` property for the `panel.background` characteristic is of type `chr` for character, and is assigned the value of `"gray92"`. How to assign a new background color?

With `ggplot2`, enclose the specific setting of the theme component, or element, in the corresponding `element` function: either `element_text()`, `element_rect()`, or `element_line()`, or, to remove an attribute, `element_blank()`. These `element` functions correspond to the basic structures of the property to be modified. The function for the `fill` property is `element_rect()`. Enclose the property names and assigned values in the `element` function call, as illustrated in the preceding code example.

style() function, **lessR**: Customize properties of visualization.

style(): Return all settings to the default theme.

theme() function, **ggplot2**: Customize properties of visualization.

10.4.3 Update and Save a Persistent Theme

Pre-defined themes can be modified to become the applicable theme for all subsequent visualizations subject to further modification.

R Input *Theme update*

```
lessR: style(panel_fill="slategray1", axis_text_color="darkred")
ggplot2: theme_update(
           panel.background = element_rect(fill="slategray1"),
           axis.text = element_text(color="darkred"))
```

For `lessR`, the `style()` function can both change the theme to a new pre-defined style and update the current style. Indicate a change to a new style by including the name of the style as the first argument of the call to the `style()` function. For ggplot2, update a theme with the `theme_update()` function.

For later use, store themes, including customized themes, as an R object. Retrieve the theme later for use in the current session, or, save as a file and then retrieve for later R sessions.

R Input *Save the current theme for later use in the same session*

```
lessR: mystyle <- style(get=TRUE)
          ... other visualizations and analyses
       style(set=mystyle)
ggplot2: mystyle <- theme_get()
          ... other visualizations and analyses
       theme_set(mystyle)
```

For `lessR`, set the `get` parameter to `TRUE` for the `style()` function to get the current theme into an R object of any valid R name. Then later retrieve the settings from the saved object by setting `set` to `TRUE`. For `ggplot2`, the `theme_get()` function saves the current theme into the designated object.

For both visualization systems the created themes are instances of a list R data structure. Here create a grayscale with a sub-theme of `wsj`. Then save for use at a later session when the new theme is re-read.

R Input *Save themes to the file system and retrieve*

```
lessR: grayWSJ <- style("gray", sub_theme="wsj")
          Write("grayWSJ", data=grayWSJ, format="R")
          ...
          mystyle <- Read("grayWSJ.rda")
          style(set=mystyle)
```

The `lessR` function `Write()` writes the R objects to a file in native R format, the current working directory, indicated in the output of the `Write()` function. The R file type is `.rda`. Then later read the file with the `lessR` function `Read()`, also presumed here in the same directory as there is no indicated path name in the file specification. Then activate `lessR` or `ggplot2` with the newly read theme, `mystyle`, as the current theme for subsequent visualizations.

10.4.4   Custom Margins

Both `lessR` and `ggplot2` calculate the placement of the axes and the axis labels dependent on the characteristics of a specific visualization. While generally appropriate, there may be some circumstances to prefer a different placement. Both systems provide for this customization.

For `lessR`, invoke the `lab_adj` and `margin_adj` parameters as part of function calls to `BarChart()`, `Histogram()`, and `Plot()`. The parameter `lab_adj` is a two-element vector for the x-axis label and y-axis label, which adjusts the position of the specified axis labels. The `margin_adj` parameter is a four-element vector – top, right, bottom and left – that adjusts the specified margins of the plot. In this example, the bottom margin moves up 1/4 inch from its initial position.

lessR: ... `margin_adj=c(0,0,.25,0)` ...

For `lessR`, positive values (+) of `lab_adj` and `margin_adj` move the initially computed corresponding margin away from plot edge in inches, and negative values (-) move that margin closer to the plot edge. For `ggplot2`, the parameter `plot.margin` specifies the margin around the plot, a `theme` element, so use for the general `theme()` function or its variants such as `theme_update()`. The value of `plot.margin` is a four element vector according to top, right, bottom, and left margins called from the `margin` function, as shown next for the specification of a smaller bottom margin than the default.

ggplot2: ... `theme(plot.margin=margin(5.5,5.5,2,5.5))` ...

The default `ggplot2` unit of measurement is the `pt`, 72 pts per inch, with default values for all four margins of 5.5pt. To apply another unit of measurement, add the unit as the fifth argument to `margin` with the appropriate abbreviation, such as `"in"` for inches. The `ggplot2` reference point is the plot edge.

Additional related `lessR` parameters are the `scale_x` and `scale_y`. Express the value of each of these parameters as a vector of three numbers: the starting value of the respective axis, the ending value, and the number of intervals. These parameters provide full control of the labeling of an axis, as well as how far the axis extends, perhaps beyond the default values.

The `lessR` parameters `pad_x` and `pad_y` provide an additional extension of the corresponding axis beyond the default value. Here the metric is different from the `scale_x` and `scale_y` parameters. Express each of the these parameters as a single value, a proportion of the original axis by which to extend the axis. The use of these parameters does not change the starting values, and the labeled intervals along the axis continues to be set by the standard R axis labeling procedure.

In summary, virtually every aspect of a visualization can be customized, either for a single visualization, or a modification that applies to subsequent visualizations as well.

References

Almende B.V., Thieurmel, B., & Robert, T. (2018). visnetwork: Network visualization using 'vis.js' library [Computer software manual]. Retrieved from `https://CRAN.R-project.org/package=visNetwork` (R package version 2.0.5)

Arnold, J. B. (2018). ggthemes: Extra themes, scales and geoms for 'ggplot2' [Computer software manual]. (R package version 4.0.1)

Bache, S. M., & Wickham, H. (2014). magrittr: A forward-pipe operator for R [Computer software manual]. (R package version 1.5)

Bar-Joseph, Z., Gifford, D. K., & Jaakkola, T. S. (2001, June). Fast optimal leaf ordering for hierarchical clustering. *Bioinformatics*, *17*(suppl 1), S22-S29.

Belsley, D. A., Kuh, E., & Welsch, R. H. (1980). *Regression diagnostics*. New York: Wiley.

Butts, C. T. (2008). network: a package for managing relational data in R. *Journal of Statistical Software*, *24*(2). Retrieved from `http://www.jstatsoft.org/v24/i02/paper`

Christie, R., & Geis, F. (1970). *Studies in Machiavellianism*. New York: Academic Press.

Cleveland, W. S. (1993). *Visualizing data*. Summit, NJ: Hobart Press.

Cohen, A. (1980). On the graphical display of the significant components in a two-way contingency table. *Communications in Statistics—Theory and Methods*, *A9*, 1025–1041.

Cohen, J. (1988). *Statistical power analysis for the behavioral sciences* (2nd ed.). Hillsdale, NJ: Lawrence Erlbaum.

Csardi, G., & Nepusz, T. (2006). The igraph software package for complex network research. *InterJournal, Complex Systems*, 1695. Retrieved from `http://igraph.org`

Friendly, M. (1994). Mosaic displays for multi-way contingency tables. *Journal of the American Statistical Association*, *89*, 190-200.

Friendly, M. (2002). Corrgrams: Exploratory displays for correlation matrices. *The American Statistician*, *56*(4), 316-324. Retrieved from `https://doi.org/10.1198/000313002533` doi: 10.1198/000313002533

Friendly, M., & Denis, D. (2005). The early origins and development of the scatterplot. *Journal of the History of the Behavioral Sciences*, *41*(2), 103-130.

Garnier, S. (2018). viridislite: Default color maps from matplotlib (lite version) [Computer software manual]. (R package version 0.3.0)

Gerbing, D. W. (2020). *The integrated violin-box-scatter (VBS) plot for the distribution of a continuous variable*. (submitted for publication)

Gerbing, D. W., & Anderson, J. C. (1988, May). An updated paradigm for scale development incorporating unidimensionality and its assessment. *Journal of Marketing Research*, *25*(2), 186-192.

Gini, C. (1921). Measurement of inequality of incomes. *The Economic Journal*, *31*(121), 124-126.

Gruvaeus, G., & Wainer, H. (1972). Two additions to hierarchical cluster analysis. *The British Journal of Mathematical and Statistical Psychology*, *25*(1), 200-206.

Hahsler, M., Hornik, K., & Buchta, C. (2008, March). Getting things in order: An introduction to the R package seriation. *Journal of Statistical Software*, *25*(3), 1-34. Retrieved from `http://www.jstatsoft.org/v25/i03/`

Hartigan, J. A., & Kleiner, B. (1984). A mosaic of television ratings. *The American Statistician*, *38*, 32-35.

Hintze, J. L., & Nelson, R. D. (1998). Violin plots: A box plot-density trace synergism. *The American Statistician*, *52*(2), 181-184.

Hubert, M., & Vandervieren, E. (2008). An adjusted boxplot for skewed distributions. *Computational Statistics and Data Analysis*, *52*, 5186–5201.

Hunter, J. E. (1973). Methods of reordering the correlation matrix to facilitate visual inspection and preliminary cluster analysis. *Journal of Educational Measurement*, *10*, 51-61.

Hunter, J. E., Gerbing, D. W., & Boster, F. J. (1982). Machiavellian beliefs and personality: Construct invalidity of the Machiavellian dimension. *Journal of Personality and Social Psychology*, *43*(6), 1293-1305.

Hyndman, R. J., & Athanasopoulos, G. (2018). *Statistical power analysis for the behavioral sciences* (2nd ed.). Melbourne, Australia: OTexts.com/fpp2. Retrieved from `https://otexts.com/fpp2/`

Hyndman, R. J., & Khandakar, Y. (2008). Automatic time series forecasting: The forecast package for R. *Journal of Statistical Software*, *26*(3), 1–22. Retrieved from `http://www.jstatsoft.org/article/view/v027i03`

Ihaka, R., Murrell, P., Hornik, K., Fisher, J. C., & Zeileis, A. (2016). colorspace: Color space manipulation [Computer software manual]. (R package version 1.3-2)

Jackson, S., Cimentada, J., & Ruiz, E. (2019). corrr: Correlations in r [Computer software manual]. Retrieved from `https://CRAN.R-project.org/package=corrr` (R package version 0.3.1)

Liiv, I. (2010, April). Seriation and matrix reordering methods. *Statistical Analysis and Data Mining*, *3*(2), 71-91.

Lock, R. H. (1993). 1993 new car data. *Journal of Statistics Education*, *1*(1).

Ludecke, D. (2018). sjplot: Data visualization for statistics in social science [Computer software manual]. Retrieved from `https://CRAN.R-project.org/package=sjPlot` (R package version 2.6.2) doi: 10.5281/zenodo.1308157

Maechler, M., Rousseeuw, P., Croux, C., Todorov, V., Ruckstuhl, A., Salibian-Barrera, M., ... Anna di Palma, M. (2016). robustbase: Basic robust statistics [Computer software manual]. (R package version 0.92-7)

Meyer, D., Zeileis, A., & Hornik, K. (2003). *Visualizing independence using extended association plots* (Vol. ISSN 1609-395X).

Murdoch, D., & Chow, E. D. (2018). ellipse: Functions for drawing ellipses and ellipse-like confidence regions [Computer software manual]. (R package version 0.4.1)

O'Hara-Wild, M. (2019). *Tidy time series forecasting with fable*. Retrieved 2/20/2019, from `https://github.com/tidyverts/fable`

Osborne, J. W., & Overbay, A. (2004). The power of outliers (and why researchers should always check for them). *Practical Assessment, Research and Evaluation, 9*(6).

Pedersen, T. L. (2018). ggraph: An implementation of grammar of graphics for graphs and networks [Computer software manual]. Retrieved from `https://CRAN.R-project.org/package=ggraph` (R package version 1.0.2)

Pedersen, T. L. (2019). tidygraph: A tidy api for graph manipulation [Computer software manual]. Retrieved from `https://CRAN.R-project.org/package=tidygraph` (R package version 1.1.2)

PROJ contributors. (2018). PROJ coordinate transformation software library [Computer software manual]. Retrieved from `https://proj4.org/`

R Core Team. (2019). R: A language and environment for statistical computing [Computer software manual]. Vienna, Austria. Retrieved from `https://www.R-project.org/`

Ram, K., & Wickham, H. (2018). wesanderson: A Wes Anderson palette generator [Computer software manual]. (R package version 0.3.6)

Revelle, W. (2018). psych: Procedures for psychological, psychometric, and personality research [Computer software manual]. Evanston, Illinois. Retrieved from `https://CRAN.R-project.org/package=psych` (R package version 1.8.10)

Rudis, B., & Gandy, D. (2017). waffle: Create waffle chart visualizations in R [Computer software manual]. (R package version 0.7.0)

Rudis, B., Ross, N., & Garnier, S. (2018). *The viridis color palettes.* Retrieved from `https://CRAN.R-project.org/package=viridis`

Sarkar, D. (2008). *Lattice: Multivariate data visualization with R.* Springer, New York, NY.

Schloerke, B., Crowley, J., Cook, D., Briatte, F., Marbach, M., Thoen, E., ... Larmarange, J. (2018). Ggally: Extension to 'ggplot2' [Computer software manual]. Retrieved from `https://CRAN.R-project.org/package=GGally` (R package version 1.4.0)

Schoonover, E. (2017). *Solarized color themes.* Retrieved from `https://ethanschoonover.com/solarized/`

Sheather, S. J., & Jones, M. C. (1991). A reliable data-based bandwidth selection method for kernel density estimation. *Journal of the Royal Statistical Society, B*, 683-690.

Sievert, C. (2018). plotly for r [Computer software manual]. Retrieved from `https://plotly-book.cpsievert.me`

Silverman, B. W. (1986). *Density estimation.* Chapman and Hall, Boca Raton, FL.

Smith, N., & van der Walt, S. (2015). A better default colormap for MatPlotLib. *Proceedings of Scientific Computing with Python.* Austin, TX, July 6–12. Retrieved from `https://matplotlib.org/`

Spence, I., & Lewandowsky, S. (1991). Displaying proportions and percentages. *Applied Cognitive Psychology, 5*(1), 61-77.

Team, R. (2016). RStudio development environment for R [Computer software manual]. Boston, MA. Retrieved from `http://www.rstudio.com/`

Tennekes, M. (2017). treemap: Treemap visualization [Computer software manual]. (R package version 2.4-2)

Tukey, J. W. (1977). *Exploratory data analysis.* Addison-Wesley, Boston, MA.

US Census Bureau, A. C. S. (2017). *Gini houshold index of income inequality.* Retrieved from `https://www.census.gov/library/publications/2017/acs/acsbr16-02.html`

Vaidyanathan, R., Xie, Y., Allaire, J., Cheng, J., & Russell, K. (2018). htmlwidgets: Html widgets for R [Computer software manual]. Retrieved from `https://CRAN.R-project.org/package=htmlwidgets` (R package version 1.3)

Vandervieren, E., & Hubert, M. (2004). An adjusted boxplot for skewed distributions. In J. Antoch (Ed.), *Proceedings in computational statistics* (pp. 1933–1940). Heidelberg: Springer-Verlag.

Wickham, H. (2007). Reshaping data with the reshape package. *Journal of Statistical Software, 21*(12), 1–20.

Wickham, H. (2014). Tidy data. *Journal of Statistical Software, 59*(10), 1–23.

Wickham, H., & Henry, L. (2019). tidyr: Tidy messy data [Computer software manual]. Retrieved from `https://CRAN.R-project.org/package=tidyr` (R package version 1.0.0)

Wilkinson, L. (2005). *The grammar of graphics.* Springer, New York, NY.

Wright, K. (2018). corrgram: Plot a correlogram [Computer software manual]. Retrieved from `https://CRAN.R-project.org/package=corrgram` (R package version 1.13)

Index

235

9 781138 599635